FUNDAMENTAL CHEMICAL KINETICS

An Explanatory Introduction to the Concepts

"Talking of education, people have now a-days" (said he) "got a strange opinion that every thing should be taught by lectures. Now, I cannot see that lectures can do so much good as reading the books from which the lectures are taken. I know nothing that can be best taught by lectures, except where experiments are to be shewn. You may teach chymestry by lectures — You might teach making of shoes by lectures!"

James Boswell: *Life of Samuel Johnson, 1766*

ABOUT THE AUTHOR

Margaret Robson Wright graduated with first class honours in Chemistry from the University of Glasgow in 1960, and then moved to Oxford University to continue her interests in chemical kinetics under the, then, Mr R.P. Bell, FRS, graduating with a D. Phil in 1963. While in Oxford she embarked on her teaching career by tutoring in Chemistry for her college, Somerville. She then moved as a Research Fellow to the Department of Chemistry of Queen's College, Dundee, then part of the University of St. Andrews. After two years she won the prestigious Shell Research Fellowship, being the first person, along with the late Dr. M.A.A. Clyne, to be awarded this honour. During her tenure of these two fellowships she continued her teaching career, lecturing, tutoring and demonstrating at all levels of the four year Scottish degree.

In 1965, she married a colleague, Dr. Patrick Wright, a fellow physical chemist with a distinguished research output who also shared her commitment to excellence in teaching. After the birth of their first child in 1967, she became a part-time lecturer concentrating totally on teaching. She held this post until 1982 when, as a result of the first set of university "cuts", she was forced, along with other part-time lecturers, into voluntary redundancy.

She then became the first University Fellow at Dundee University, an honorary position awarded to academic staff who, during their time in the university, performed outstanding teaching or research. During her seven years in the wilderness of unemployment, she maintained her interests in teaching through tutoring S-grade, H-grade and A-level school pupils and at one stage prepared students for the Cambridge University Entrance Examination. During this period she also wrote her first book *The Nature of Electrolyte Solutions* published in 1988.

In 1989 she became part-time temporary lecturer in chemistry at the University of St. Andrews, becoming part-time Teaching Fellow in Chemistry in 1992, a post she held until her retirement.

During this time she lectured and tutored on most of the topics in mainstream physical chemistry, covering all aspects of kinetics, thermodynamics, spectroscopy, equilibria and electrolyte solutions at all levels of the four year Scottish degree. She has written a wide variety of informal study guides over a considerable range of topics, produced problem booklets with extensive and detailed explanatory answers and a series of booklets on help in mathematics for chemists with a weak background in mathematics. Being solely available at St Andrews, these were voted as being very popular and extremely helpful to students.

After retirement she published two further books, *An Introduction to Chemical Kinetics* and *An Introduction to Aqueous Electrolyte Solutions*.

It is with this wide range of teaching experience that she has written this text, which reflects her total commitment to the teaching of university students.

Margaret Wright is married with two sons and a daughter, all now grown up.

FUNDAMENTAL CHEMICAL KINETICS

An Explanatory Introduction to the Concepts

Margaret Robson Wright, BSc, DPhil
formerly Lecturer in Chemistry,
The School Of Chemistry
St. Andrews University, Scotland

WOODHEAD
PUBLISHING

Oxford   Cambridge   Philadelphia   New Delhi

Published by Woodhead Publishing Limited,
80 High Street, Sawston, Cambridge CB22 3HJ, UK
www.woodheadpublishing.com

Woodhead Publishing, 1518 Walnut Street, Suite 1100, Philadelphia,
PA 19102-3406, USA

Woodhead Publishing India Private Limited, G-2, Vardaan House, 7/28 Ansari Road,
Daryaganj, New Delhi – 110002, India
www.woodheadpublishingindia.com

First published by Horwood Publishing Limited, 1999
Reprinted by Woodhead Publishing Limited, 2011

British Library Cataloguing in Publication Data
A catalogue record for this book is available from the British Library

ISBN 978-1-898563-60-0

DEDICATION

Dedicated to the memory of
my mother, Anne,
with much love, affection and deep gratitude

and to
my husband Patrick:
Anne, Edward, Andrew and the cats.

PREFACE

This book contains many important topics in modern theoretical studies in chemical kinetics, and discusses some of the advances in experimental techniques which have enabled these theories to be tested and validated.

As the title indicates the primary aim is to help students to understand modern thinking about "how a chemical reaction occurs" by giving a slow, explanatory 'student-friendly' yet rigorous introduction to such thinking. Many students veer rapidly away from topics which can be labelled theoretical or conceptual because of a near involuntary and automatic assumption that they will be incapable of understanding. This book attempts to allay these fears by guiding the student through these modern topics in a step-by-step development which explains both the logic and the reasoning.

There are two major problems for students to overcome before they can understand a theory: firstly, the concepts, and then the mathematics used to quantify them. In the author's experience students rarely manage to deduce the chemical concepts from the mathematical development. These concepts need to be fully and explicitly explained first before the student can start to understand 'how chemical reaction occurs'. It is the author's firm belief, based on years of teaching, that students can gain a considerable understanding and knowledge of what is involved in a theoretical discussion of chemical reaction without delving into the mathematics first. Once the concepts have been assimilated it is then time to move on to the mathematical presentation, which in this book is developed with the same full explanatory detail.

It is unfair to expect students to work all this out for themselves, over and above the very necessary learning of facts. It is the teacher's duty to show them how to achieve understanding and to explain how to think scientifically. The philosophy behind this book is that this is best done by detailed explanation and guidance. It is understanding and confidence which help to stimulate and sustain interest. This book attempts to do precisely that.

For a subject so wide-ranging as chemical kinetics, and in a book of this length, it is impossible to cover all theoretical topics, especially if explanation is the key aim. For this reason, the book is not comprehensive, and only the more important topics are covered.

ACKNOWLEDGEMENTS

This book is the result of the accumulated experience of nearly forty very stimulating years of teaching students at all levels. During these years I learned that being happy to help, being prepared to give extra explanation and to spend extra time on a topic could soon clear up problems and difficulties which many students thought they would never understand. Too often teachers forget that there were times when they themselves could not understand, and when a similar explanation and preparedness to give time were welcome. If, through the written word, I can help students to understand and to feel confident in their ability to learn, and to teach them in a book in a manner which gives them the feeling of direct contact with the teacher, then this book will not have been written in vain. To all the many students who have provided the stimulus and enjoyment of teaching I give my grateful thanks.

I would also like to thank Ellis Horwood of Horwood Publishing for his encouragement to present this book for publication, and for his belief in the method of approach. His continuing encouragement, enthusiasm and help in preparing the manuscript is very gratefully acknowledged and appreciated.

My thanks are due to Mrs Jill Blyth for a first-rate rendering of the hand-written manuscript into a working typescript.

My especial and very grateful thanks are given to Mrs Rosmary Harris of Rosetec, Worthing, who made the publication of this book possible by her expert and speedy conversion of the initial typescript into camera ready form. Her help with the technical aspects was vital and indispensible, and I have been totally dependent on her skills. My thanks are also due to Mrs Shirley Fox of the Department of Mathematics in the University of Dundee who typed the mathematical parts of the book.

To my mother, Mrs Anne Robson, I have a very deep sense of gratitude for all the help she gave me in her lifetime in furthering my academic career. I owe her an enormous debt for her invaluable, excellent and irreplaceable help with my children when they were young and I was working part-time during the teaching terms of the academic year. Without her help and her loving care of my children I would never have gained my continued experience in teaching, and I would never have written this book. My deep and most grateful thanks are due to her.

My husband, Patrick, has throughout my teaching career and throughout the thinking about and writing of this book, been a source of constant support, help and encouragement. His very high intellectual calibre and wide-ranging knowledge and understanding have provided many fruitful and interesting discussions. He has been an excellent sounding board for many of the ideas and manner of presentation in this book. He has read in detail the whole manuscript and his clarity, insight and considerable knowledge of the subject matter has been of invaluable help. My debt to him is enormous, and my most grateful thanks are due to him.

Finally, my thanks are due to my three children who have encouraged me throughout my career and, in particular, encouraged me to write this book. Their very fine and keen brains have always been a source of stimulus to me, and have always kept my interest in young people flourishing.

Margaret Robson Wright
School of Chemistry
The University of St Andrews
January, 1999

CONTENTS

1

Introduction:
Historical Developments and Modern Kinetics

1.1 A historical development of concepts in kinetics

The history and development of chemical kinetics gives a fascinating insight into the way in which theory and experiment can become mutually interdependent. Experimental work has been steadily maintained since about 1850, interspersed with periods of intense theoretical activity. Both aspects have become more and more important as fundamental features of general chemistry, and both have progressed dramatically in techniques and in power in the period since 1950.

Arrhenius laid the foundations of the subject as a rigorous science when he postulated that not all molecules can react: only those reactant molecules which have a certain minimum energy, called the activation energy, can react. The fraction of reactant molecules which possess this critical energy is calculated from the Maxwell-Boltzmann distribution which is the keystone of most theoretical approaches. These early theoretical developments encouraged more experimental work which was then used as a testing ground for the next major theoretical development, collision theory, proposed around the early part of this century. This was worked out for a bimolecular reaction where it was assumed that reaction could only occur when two reactant molecules collide. Kinetic theory then enabled an expression for the rate of collision to be derived. When this was compared with observed rates, the theoretical expression was generally found to be much too high. However, a rate of reaction can still be calculated by making the assumption that only a certain fraction of the molecules which collide will actually be able to react, and this fraction is the number of molecules which have energy above the critical minimum value. This modified collision theory met with considerable success for a large number of reactions, though there were still some reactions where the calculated rate was too high. To account for this the steric factor,"p", was introduced. However, this remained an empirical factor limiting the collision theory rate, and no theoretical status was given.

Unfortunately, collision theory appeared to be unable to account for unimolecular reactions which showed first order behaviour at high pressures, and second order behaviour at low pressures. If the determining factor in the rate is the rate at which molecules collide, unimolecular reactions might be expected always to give second order kinetics, which is not what is observed.

For some time there was considerable confusion about unimolecular reactions which appeared to be emerging as a class on their own, and it was not until Lindemann formulated growing speculation into the now famous Lindemann-Christiansen hypothesis that a theoretical description of unimolecular reactions was developed. The mechanistic framework used since then in all more sophisticated unimolecular theories is also common to both bimolecular and termolecular reactions.

The crucial argument is that molecules which are activated and have the required minimum activation energy do not have to react immediately they receive this energy by collision. There is sufficient time after the final activating collision for the molecule either to be deactivated in another collision, or to react in a unimolecular step.

It is the existence of this time lag between activation by collision and reaction which is basic and crucial to the theory. This assumption leads inevitably to first order kinetics at high pressures, and second order kinetics at low pressures.

Other elementary reactions can be handled in the same fundamental way; molecules can be activated by a binary collision and then last long enough during the time lag for there to be the same two fates open to them. The only difference lies in the molecularity of the reaction step,

1) in a unimolecular reaction, only one molecule is involved in the moment of chemical reaction
2) in a bimolecular reaction, two molecules are involved in this step
3) in a termolecular reaction, three molecules are now involved.

Meanwhile the idea of reaction being defined in terms of the spatial arrangement of all the atoms in the reacting system was being explored, and in the late 1920's transition state theory was finally presented. This theory was to prove to be of fundamental importance. Reaction is now defined as the acquisition of a certain critical configuration of all the atoms involved in the reaction, and this critical configuration was shown to have a critical maximum in potential energy with respect to the reactants and products.

The lowest potential energy pathway between reactant and product configurations represents the changes which take place during reaction. The critical configuration lies on this pathway at the configuration with the highest potential energy. It is called the activated complex or transition state, and it must be attained before reaction can occur. The rate of reaction is the rate of change of configuration through the critical configuration. This theory will be developed in Chapters 2 and 3.

Transition state theory has proved to be a very powerful tool, vastly superior to collision theory, and only recently challenged by the modern advances in molecular beams and molecular dynamics which look at the microscopic details of a collision, and which can be regarded as a modified collision theory. These topics are developed in Chapters 7 to 10.

The development of the experimental technique of molecular beams, and the computational technique of molecular dynamics have revolutionised the science of reaction rates. Kinetics has now moved to an even more fundamental level of study, where it is probing at the very heart of the microscopic details of molecular behaviour and reaction. There are now two branches of kinetics, conventional and microscopic studies, each of which alone constitutes a major aspect of chemistry, but which fully complement each other.

Experimental techniques have also made astounding advances throughout the latter half of this century, and these are described in Chapter 7 to 10. Experimental techniques have been forced to advance in order to meet the two complementary needs (a) of being a tool with which to test theoretical advances, and (b) acting as an inspiration to develop theory to account for the experimental results.

For a long time chemists have used the term "mechanism" to refer to the chemical steps which make up a reaction. There are some reactions which occur via one chemical step only, and these are called elementary, but the vast majority of reactions proceed by two or more steps. However, in the last decade or so there has been a vast upsurge of interest in the physical mechanism of reaction. This is a description at an even more fundamental microscopic level than the molecular level of the chemical mechanism. A whole new field of kinetics has developed out of this interest, and the great advances in molecular beam and molecular dynamic studies reflect this interest.

1.2 Concepts involved in modern state to state kinetics

State to state kinetics involves looking at the behaviour of individual molecules. These can be characterised by their quantum states which summarise information about the translational, rotational, vibrational and electronic energy of each molecule. Modern work looks at the processes of energy transfer which put molecules into these quantum states. It also looks at the details of reaction from these quantum states, and determines the quantum states of the products. Rate constants for these energy transfer processes and for reaction between reactants and products each in specific quantum states are determined, and these are called state to state rate constants.

This modern work relies heavily on developments in two main areas. Molecular beam experiments study the details of collisional energy transfer between specific quantum states, and give rates of reaction between molecules in specific quantum states. They aim at a complete physical description of what happens in a single collision, whether reactive or non-reactive. This results in a very clear description of the physical processes and mechanisms involved.

Laser-induced fluorescence developments have been a crucial factor in the study of state to state processes. Laser induced fluorescence allows the quantum states and their populations, even though very low, to be found for the molecules being studied in molecular beam experiments. Without laser induced fluorescence many of the state to state experiments could not be carried out.

Development of fast, large memory computers has enabled accurate quantum calculations on many small molecules to be made. Quantum mechanical potential energy surfaces result, and details of translational, rotational and vibrational energies can be superimposed on these surfaces giving valuable information about reactive and non-reactive collisions.

Trajectory calculations then allow predictions to be made about state to state processes of non-reactive and reactive collisions. These calculations can then be compared with observed behaviour in molecular beam experiments.

Details of these modern techniques are given in Chapters 7 to 10.

1.3 Collision processes and a master mechanism describing elementary reactions at the microscopic level

Once the molecular chemical mechanism is known for any reaction, a detailed microscopic description can then be given to the moment of chemical transformation and to the fundamental physical processes occurring during reaction. All reactions, no matter how complex, are made up of a series of elementary processes, and it is specifically these individual steps which must be described at the microscopic level. There are three important features of the kinetic aspects.

Collisional processes are fundamental aspects of all reactions, and they play a dual role in kinetics. They are the means whereby energy is exchanged between molecules, and possibly also within the molecules themselves. Collisions are, therefore, fundamental to all activation and deactivation steps. They may also be important in the reaction step. A collision is necessary in bimolecular and termolecular reaction steps, but is not needed in the spontaneous breakdown of a unimolecular reaction step.

Molecules can only react once they have acquired a certain critical minimum energy, but they do not react *immediately* they get this energy. This requires that activated molecules have a finite lifetime during which they can either be deactivated or react.

The reaction step is characterised by a series of configurational changes which involve an internal rearrangement of the relative positions of all the atoms involved in the reaction step.

These points are summarised in a mechanism common to all elementary reactions irrespective of molecularity.

$$A + A \quad \xrightarrow{k_1} \quad A^* + A \quad \text{bimolecular activation by collision}$$

$$A^* + A \quad \xrightarrow{k_{-1}} \quad A + A \quad \text{bimolecular deactivation by collision}$$

$$A + bA^* \xrightarrow{k_2}$$ products: the reaction step which involves configuration changes taking the molecule through the critical configuration of the activated complex

A^* is an activated molecule with enough energy to react.

$b = 0$ defines spontaneous unimolecular breakdown of A^*
$b = 1$ defines bimolecular reaction involving the coming together of A^* with A
$b = 2$ defines termolecular reaction involving the coming together of A^* with two A's

1.3.1 Energy transfer processes in the master mechanism

The activation/deactivation steps in the master mechanism are simply energy transfer processes, three aspects of which are of fundamental significance to the theoretical kineticist,

i)   the type of energy involved in the accumulation of energy,
ii)   the rate of accumulation of this energy,
iii)   the physical mechanism of energy transfer.

The total energy of a molecule has contributions from

i)   the translational energy of the molecule in space,
ii)   the rotational and vibrational energies,
iii) the electronic energy,

and the total energy is distributed among these types of energy according to the Maxwell-Boltzmann distribution. For a non-reacting gas at constant temperature, the number of molecules possessing a given type of energy is fixed, as is the distribution of these energies. But this is an *overall* situation. Although the overall distribution is fixed, and the overall population of any state is fixed, the particular molecules possessing a given energy are not fixed. Molecules can still exchange energy between each other, and do so subject only to the condition that the amounts and kind of energy exchanged must be such that the overall distribution of energy is kept fixed. And so translational energy in one molecule can be converted to vibrational energy in another molecule, and vice-versa, or electronic energy in one molecule can be converted into vibrational and rotational energy in another.

1.3.2 Accumulation of energy in kinetics
Three questions can be asked:

i)   what types of energy can be counted towards accumulation of activation energy?
ii)   are different types of energy required for different types of elementary reaction?
iii) once the activation energy has been accumulated, does there have to be a
      redistribution of that energy?

These questions are normally only raised in the context of unimolecular reactions, but all three are fundamental features in any discussion of reaction, irrespective of molecularity. It is only modern work such as state to state kinetics which emphasises the relevance of these quantities for all molecularities.

Accumulation of translational energy is the simplest energy transfer and is also the easiest. However, it is rather naive to assume that the violence of a collision is the only condition for activation. Molecules have internal motions corresponding to vibrations and rotations, and these could contribute to producing conditions favourable to reaction. A more violently vibrating molecule could be more likely to react than a molecule in its ground vibrational state. Furthermore, the violence of a collision can affect the internal states of the molecule, and by altering the vibrational or rotational energy of a molecule can affect its chances of reacting in a subsequent collision.

Accumulation of vibrational energy is a much more complex situation. The number of vibrational modes in a molecule increases rapidly as the number of atoms in the molecule increases, Table 1.1. So even a simple molecule has a significant number of modes, some or all of which could be utilised in activation.

Table 1.1 Total number of vibrational modes for non-linear molecules with differing numbers of atoms

Number of Atoms	Example of Molecule	Total number of vibrational modes possible for a non-linear molecule = $3N$-6 where N is the number of atoms in the molecule
2	Cl_2 (linear)	1
4	CF_3'	6
5	CH_4	9
6	$CH_2 = CH_2$	12
8	C_2H_6	18
11	C_3H_8	27
32	$C_{10}H_{22}$	90

The contribution from rotational energy is likely to be small because the number of modes is small. But this does not mean that rotational energy is unimportant and can be ignored. Modern experimental results from ultrasonics, molecular beams, laser induced fluorescence and chemiluminescence all suggest that rotational states as well as vibrational states should be considered.

In unimolecular reactions the main emphasis has been on accumulation of vibrational energy. The greater the vibrational energy of a molecule the more likely it is to break down spontaneously into products. In principle, this could be checked with laser experiments which place the molecule in specific high vibrational states from which reaction can occur. The rate of reaction from these specific quantum states will show whether there is a dependence on vibrational energy. Rotational energy is not normally considered, but modern work suggests that contributions from rotational energy should be included.

Accumulation of translational energy and the violence of a collision have been considered to be the most important features for activation in bi and termolecular reactions. Reaction occurs by a series of configurational changes involving internal changes, and therefore vibration and rotation should be considered for these molecularities. Modern developments allow for this possibility, and look at rates of reaction from specific vibrational and rotational states. Molecular beam experiments have shown that for some bimolecular reactions a critical minimum vibrational energy is needed.

1.3.3 Redistribution of accumulated energy

Once molecules have acquired their critical energy, they are potential reacting systems. However, it may be that redistribution of the critical energy has to occur after activation, but before reaction. If this is so, details of *intra* molecular energy transfer will be required, and these transfers will occur in the time interval *between* collisions.

Translational energy cannot be converted into vibrational or rotational energy in the *same* molecule. In contrast, vibrational energy can be redistributed around the various normal modes in an intramolecular process. Alteration in rotational states often accompanies such vibrational energy redistribution.

Unimolecular theory assumes that sufficient energy must be accumulated in the required number of modes, but reaction cannot occur until this energy is suitably distributed among the total number of vibrational modes. Such a critical redistribution of vibrational energy is not considered in bimolecular reactions, but again molecular beam experiments force a reassessment of this.

1.3.4 Rate and mechanism of accumulation of energy
Under certain conditions the rate of accumulation of energy could determine the rate of reaction. The calculations involved are extremely difficult, and despite recent advances, are still incomplete.

All conceivable energy transfers must be included in this calculation. This means that rate constants are required for transfers such as vibrational to vibrational energy, rotational to vibrational energy, rotational to rotational energy as well as transfer of translational energy to translation, rotation and vibration. Transfers of electronic energy are not considered when reaction occurs in the ground electronic state throughout. Molecular beam experiments coupled with laser induced fluorescence are a major source of information.

The mechanism of energy transfer is also highly pertinent. Transfer of kinetic energy is very easy and occurs on every collision. Transfer of translation to rotation is also fairly easy, often requiring only a few collisions, but transfer to vibration is a very difficult process sometimes requiring thousands of collisions. Transfers such as vibration to vibration or rotation can be easy or difficult, and this depends on whether the transfer is resonant or not (Chapter 9).

Activation and deactivation are energy transfer processes. Activation could progressively excite a molecule to higher vibrational or rotational states by single or small quantum jumps such as

$$0 \rightarrow 1, 1 \rightarrow 2, 2 \rightarrow 3 \dots$$
$$0 \rightarrow 1, 1 \rightarrow 2, 2 \rightarrow 4 \dots$$

This is called *ladder climbing*, and only small amounts of energy are exchanged. Energy exchanges could also occur by multi-quantum jumps such as

$$0 \rightarrow 4, 1 \rightarrow 7, 7 \rightarrow 18, 18 \rightarrow 30 \dots$$

This is called *strong collisions*, and large amounts of energy can be exchanged in this way. These process are discussed in detail in Chapter 8.

1.3.5 Lifetimes, intervals between collisions, and their implications for kinetics
The time between acquisition of activation energy and the moment of chemical transformation is the lifetime of an activated molecule, or the time lag.

The time taken to pass through the critical configuration is the lifetime of an activated complex.

If the time between collisions is long compared with the time lag then activated molecules will generally react before they can be deactivated. The slow steps are, therefore, activation and deactivation, and the details of energy transfer will play a central role in determining the rate of reaction, and in the development of the theoretical treatment.

If the time interval between collisions is short compared with the time lag then activated molecules will generally be deactivated before they can react. The slow step is, therefore, reaction and discussion will focus on the attainment of the critical configuration of the activated complex, with the details of energy transfer occupying an implicit role in the developing theory.

1.3.6 Possibility of equilibrium between reactants and activated molecules

In a non-reacting gas there is an equilibrium between activation and deactivation

$$A + A \; \overset{k_1}{\underset{k_{-1}}{\rightleftharpoons}} \; A^* + A$$

$$K_{eq} = \left(\frac{[A^*]}{[A]} \right)_{eq} = \frac{k_1}{k_{-1}}.$$

(1.1)

K_{eq} can be calculated from statistical mechanics using the Maxwell-Boltzmann distribution.

In a reacting gas, activated molecules are also being removed by reaction,

$$A + A \; \overset{k_1}{\rightarrow} \; A^* + A$$

$$A^* + A \; \overset{k_{-1}}{\rightarrow} \; A + A$$

$$A^* + bA \; \overset{k_2}{\rightarrow} \; products$$

so that equilibrium cannot be assumed and steady state, or still more general treatments, have to be applied to the activated molecules.

When equilibrium is established a steady equilibrium concentration of A^* is set up where

$$\frac{d[A^*]_{eq}}{dt} = 0,$$

that is, there is no overall change in the *equilibrium* concentration.

In contrast, when a steady state is set up there is a very low but steady concentration of A^* set up such that there is virtually no change in the *steady state* concentration of A^*.

$$\frac{d[A^*]_{ss}}{dt} = 0.$$

But it must be realised that this steady concentration is *not an equilibrium concentration*.

$$\frac{d[A^*]}{dt} = k_1[A]^2 - k_{-1}[A^*][A] - k_2[A^*][A]^b = 0$$

$$[A^*] = \frac{k_1[A]}{k_{-1} + k_2[A]^{b-1}}$$

(1.2)

and

$$\frac{[A^*]}{[A]} = \frac{k_1}{k_{-1} + k_2[A]^{b-1}}.$$

(1.3)

If the time interval between collisions is short compared to the time lag between activation and reaction, most A* molecules will be removed by collision rather than by reaction, and so the reaction step will not severely disturb the equilibrium concentration of activated molecules, and a Maxwell-Boltzmann distribution will probably be maintained. This corresponds to

$$k_{-1} \gg k_2[A]^{b-1}$$

(1.4)

$$k_{-1}[A^*][A] \gg k_2[A^*][A]^b.$$

(1.5)

Under these conditions a statistical mechanical calculation would give the equilibrium constant for activation/deactivation, and this will be approximately equal to [A*]/[A].

$$K_{eq} = \left(\frac{[A^*]}{[A]}\right)_{eq} = \frac{k_1}{k_{-1}}.$$

(1.6)

If, however, the time between collisions is long compared to the time lag between activation and reaction, most A* molecules will be removed by reaction and this will severely deplete the equilibrium concentration of A*, and a Maxwell-Boltzmann distribution will certainly not be set up. This corresponds to

$$k_{-1} \ll k_2[A]^{b-1}$$

(1.7)

$$k_{-1}[A^*][A] \ll k_2[A^*][A]^b$$

(1.8)

and so the ratio $[A^*]/[A]$ will not be equal to either k_1/k_{-1} or K_{eq}.

The postulate of a time lag raises the question, "what actually happens" during the time lag? This focuses attention on to the need for reacting molecules with the required critical energy to rearrange their configuration into the critical configuration required for reaction.

As will be shown later the existence of a time lag requires a unimolecular reaction to have first order kinetics at high pressures with a move to second order at low pressures. The existence of a time lag also predicts that a termolecular reaction should show a pressure dependence for the observed rate, but a bimolecular reaction should be second order at all pressures. Although the postulate of a time lag is not essential to explain the second order behaviour of bimolecular reactions, it is nonetheless a relevant feature of a general treatment of all elementary reactions whether uni, bi or termolecular. Although this point was glossed over in earlier theoretical treatments it has moved into prominence in the microscopic nature of modern theoretical and experimental work.

1.4 Kinetics of uni-, bi-, and termolecular reactions using the master mechanism
The three mechanisms are analysed in Table 1.2 for the simple reactions:

$$A \qquad\qquad \rightarrow \qquad products$$
$$A + A \qquad\quad \rightarrow \qquad products$$
$$A + A + A \qquad \rightarrow \qquad products$$

Table 1.2

Unimolecular Mechanism	Bimolecular Mechanism	Termolecular Mechanism
I	II	III
$A + A \xrightarrow{k_1} A^* + A$	$A + A \xrightarrow{k_1} A^* + A$	$A + A \xrightarrow{k_1} A^* + A$
$A^* + A \xrightarrow{k_{-1}} A + A$	$A^* + A \xrightarrow{k_{-1}} A + A$	$A^* + A \xrightarrow{k_{-1}} A + A$
$A^* \xrightarrow{k_2}$ products	$A^* + A \xrightarrow{k_2}$ products	$A^* + A + A \xrightarrow{k_2}$ products

Applying the steady state treatment to A* in all three mechanisms:

Unimolecular Mechanism

$$\frac{d[A^*]}{dt} = k_1[A]^2 - k_{-1}[A^*][A] - k_2[A^*] = 0$$

$$[A^*] = \frac{k_1[A]^2}{k_{-1}[A]+k_2} = \frac{k_1[A]}{k_{-1}+k_2/[A]}$$

$$\text{Rate of Reaction} = k_2[A^*]$$

$$= \frac{k_1 k_2[A]^2}{k_{-1}[A]+k_2} = \frac{k_1 k_2[A]}{k_{-1}+k_2/[A]}$$

Bimolecular Mechanism

$$\frac{d[A^*]}{dt} = k_1[A]^2 - k_{-1}[A^*][A] - k_2[A^*][A] = 0$$

$$\therefore [A^*] = \frac{k_1[A]}{k_{-1}+k_2}$$

$$\text{Rate of reaction} = k_2[A^*][A]$$

$$= \frac{k_1 k_2[A]^2}{k_{-1}+k_2}$$

Termolecular Mechanism

$$\frac{d[A^*]}{dt} = k_1[A]^2 - k_{-1}[A^*][A] - k_2[A^*][A]^2 = 0$$

$$\therefore [A^*] = \frac{k_1[A]}{k_{-1}+k_2[A]}$$

$$\text{Rate of reaction} = k_2[A^*][A]^2$$

$$= \frac{k_1 k_2[A]^3}{k_{-1}+k_2[A]}$$

Order of reaction is neither first order nor second

Rate of reaction $= k_{obs}[A]$

and

$$k_{obs} = \frac{k_1 k_2 [A]}{k_{-1}[A] + k_2}$$

The observed rate constant is dependent on concentration

A* molecules have two fates

(1) They can be destroyed by deactivation
(2) They can be destroyed by reaction

The partitioning of the activated molecules A* between their two possible fates is given by the ratio:

$$\chi = \frac{\text{chance of deactivation (from A*)}}{\text{chance of conversion to the activated complex (from A*)}}$$

Order of reaction is neither first order nor second

$$\chi = \frac{k_{-1}[A^*][A]}{k_2[A^*]}$$

$$= \frac{k_{-1}[A]}{k_2}$$

Order of reaction is strictly second

Rate of reaction $= k_{obs}[A]^2$

and

$$k_{obs} = \frac{k_1 k_2}{k_{-1} + k_2}$$

The observed rate constant is independent of concentration

$$\chi = \frac{k_{-1}[A^*][A]}{k_2[A^*][A]}$$

$$= \frac{k_{-1}}{k_2}$$

Order of reaction is neither third nor second

Rate of reaction $= k_{obs}[A]^3$

and

$$k_{obs} = \frac{k_1 k_2}{k_{-1} + k_2[A]}$$

The observed rate constant is dependent on concentration

$$\chi = \frac{k_{-1}[A^*][A]}{k_2[A^*][A]^2}$$

$$= \frac{k_{-1}}{k_2[A]}$$

χ is necessarily concentration and pressure dependent	χ is a function of rate constants alone	χ is necessarily concentration and pressure dependent
Reaction is *first* order when $$k_{-1}[A] \gg k_2$$ $$k_{-1}[A][A^*] \gg k_2[A^*]$$	Reaction is always *second* order $$k_{-1} \text{ can be} > \text{ or } < k_2$$ *even if*	Reaction is *second* order when $$k_{-1} \ll k_2[A]$$ $$k_{-1}[A][A^*] \ll k_2[A]^2[A^*]$$
rate of deactivation $\gg$ rate of reaction	rate of deactivation $>$ rate of reaction or rate of deactivation $<$ rate of reaction	rate of deactivation $\ll$ rate of reaction
This happens when [A] is large i.e. at high concentrations or pressures	This happens over all concentrations and pressures	This happens when [A] is large i.e. at high concentrations or pressures
Corresponds physically to: the interval between collisions is *short* compared to the time lag	Corresponds physically to *not knowing* whether the interval between collision is *short* or *long* compared to the time lag	Corresponds physically to: the interval between collisions is *long* compared to the time lag
And means that most A* are removed by deactivation rather than reaction	And means most A* could be removed either by deactivation or by reaction; observed kinetics cannot distinguish these	And means that most A* are removed by reaction rather than deactivation

And hence reaction is rate-determining	And hence do not know which step, activation or reaction, is rate-determining	And hence activation is rate-determining
And a Maxwell-Boltzmann distribution is probably set up	End up *assuming* a Maxwell-Boltzmann distribution to be set up	And a Maxwell-Boltzmann distribution is certainly *not* set up
And so the supply of activated molecules will be *adequate*	It is assumed that the supply of activated molecules is *adequate*	And so the supply of activated molecules is *inadequate*
Reaction is *second* order when $k_{-1}[A]$ << k_2 $k_{-1}[A][A^*]$ << $k_2[A^*]$ rate of deactivation << rate of reaction		Reaction is *third* order when k_{-1} >> $k_2[A]$ $k_{-1}[A][A^*]$ >> $k_2[A]^2[A^*]$ rate of deactivation >> rate of reaction
This happens when [A] is small, that is at low concentrations and pressures		This happens when [A] is small, that is at low concentrations or pressures
Corresponds physically to: the interval between collisions is *long* compared to the time-lag		Corresponds physically to: the interval between collisions is *short* compared to the time-lag

		And means that most A* are removed by deactivation rather than reaction
		And hence reaction is rate-determining
	End up *assuming* a Maxwell-Boltzmann distribution is set up	And a Maxwell-Boltzmann distribution is probably set up
	It is assumed that the supply of activated molecule is adequate	And so the supply of activated molecules will be *adequate*
	The rate constant is strictly second order, and is independent of concentration or pressure	A change in order with concentration or pressure is observed experimentally
	And the reaction and deactivation steps have the *same* molecularities	And the reaction and deactivation steps have *different* molecularities
And means that most A* are removed by reaction rather than deactivation		
And hence activation is rate-determining		
And a Maxwell-Boltzmann distribution is *certainly not* set up		
And so the supply of activated molecules is *inadequate*		
A change in order with concentration or pressure is observed experimentally		
And the reaction and deactivation steps have *different* molecularities		

If the bimolecular reaction were

$$A + B \rightarrow C + D$$

many more steps would need to be considered for activation, deactivation and reaction. For instance, an activated A^* molecule could also be formed by collisions with B, C and D molecules.

These cases result in much more complex algebra than do the simple cases quoted, but the general results and conclusions are the same as for the simplified cases. In the table the rate expressions are given in terms of concentrations, even though the discussion is given in terms of pressures. This is a typical description of the kinetics of gas phase reactions for which

$$pV = nRT$$
$$p = cRT.$$

$$(1.9)$$

1.4.1 Comments on the table

From the table it can be seen that a bimolecular reaction is always second order, and that the observed rate constant is independent of pressure. The partitioning of the activated molecule between its two fates is also independent of pressure. It is impossible to tell whether activation or reaction is rate determining, though most theories focus attention on to the reaction step.

Uni and termolecular reactions show an order which varies with pressure, and the partitioning of the activated molecule between its two fates is also pressure dependent. For both reaction types, the time lag and the time between collisions swap over in relative magnitude between low and high pressures, resulting in the rate determining step being different at the two pressure limits.

In contrast to gas phase reactions, analogous transitions are never found for reactions in solution. Reaction is always the rate determining step since solvent molecules are able to maintain an adequate concentration of activated molecules. The only exceptions are diffusion controlled reactions.

1.5 A generalisation of these ideas

All these ideas can be taken over to describe the details of even the most complex of chemical reactions where there are several elementary individual steps, for example the chain reaction between H_2 (g) and Br_2 (g) can be explained by the following mechanism.

$Br_2 + M$	$\xrightarrow{k_1}$	$2Br\cdot + M$	initiation
$Br\cdot + H_2$	$\xrightarrow{k_2}$	$HBr + H\cdot$	propagation
$H\cdot + Br_2$	$\xrightarrow{k_3}$	$HBr + Br\cdot$	
$H\cdot + HBr$	$\xrightarrow{k_4}$	$H_2 + Br\cdot$	inhibition
$2Br\cdot + M$	$\xrightarrow{k_5}$	$Br_2 + M$	termination

The overall kinetics are complex, and the rate is given by

$$\text{rate} = \frac{k_2 (k_1/k_5)^{\frac{1}{2}} [H_2][Br_2]^{\frac{1}{2}}}{1 + (k_4/k_3)[HBr]/[Br_2]}.$$

(1.10)

At the macroscopic level this reaction looks totally different from simple elementary reactions such as

$$NO_2(g) + F_2(g) \quad \rightarrow \quad NO_2F(g) + F\cdot(g)$$

$$\text{rate} = k[NO_2][F_2]$$

(1.11)

or

$$2ClO\cdot(g) \quad \rightarrow \quad Cl_2(g) + O_2(g)$$

$$\text{rate} = k[ClO\cdot]^2.$$

(1.12)

However, at the microscopic level the H_2/Br_2 reaction is simply a series of elementary reactions as given in the reaction mechanism. Each of these steps can be described by the general mechanism.

$$A + A \quad \rightarrow \quad A* + A$$
$$A* + bA \quad \rightarrow \quad \text{products}$$

with the same conclusions as were drawn in the Table 1.2.

2

Basic Transition State Theory

It is usually accepted that molecules have to be suitably activated before they can react. Transition state theory describes what happens to these activated molecules in terms of the configurational changes which lead to reaction. Transition state theory has proved to be very successful in the interpretation of experimental results, and has only recently been challenged by modern developments in molecular dynamics. It gives a detailed calculation of the absolute rate of a chemical reaction, and the statistical mechanical derivation makes considerable use of the properties of potential energy surfaces. Although fundamentally a gas phase theory, it can be applied to solutions. Here it is less rigorous because the effect of the solvent is difficult to quantify.

Transition state theory is only applicable when the rate determining step is the reaction step, and must never be used to discuss the activation step since energy transfers are totally outside the remit of the theory.

2.1.1 Transition state theory and configuration
In the process of reaction the reacting molecules come sufficiently close together so that interactions are set up between all the atoms involved. The reacting molecules and products are described in terms of all the atoms in a single unit made up of all the reacting species and products. For example the reaction

$$CH_3\cdot + C_2H_6 \quad \rightarrow \quad CH_4 + C_2H_5\cdot$$

is described in terms of the unit

"reaction unit"

rather than as the independent species

$$CH_3\cdot, \quad C_2H_6, \quad CH_4, \quad C_2H_5\cdot$$

2.1.2 Transition state theory and potential energy
When the configuration changes, the potential energy of the "reaction unit" also changes, and a potential energy surface summarises these changes. A description of

what happens during reaction rests on a knowledge of the potential energy surface, as does the actual derivation of the theory.

The potential energy surface and its properties can be given for the general reaction

$$A + B \rightarrow C + D$$

where A, B, C and D are polyatomic molecules. This would give an n-dimensional surface. Fortunately, all the properties of an n-dimensional surface which are relevant to kinetics can be exemplified by a three dimensional surface which is much easier to visualise and to describe. A three dimensional surface describes the reaction

$$X + YZ \rightarrow XY + Z$$

where X, Y and Z are atoms, and a linear approach of X to YZ and recession of Z is assumed, implying a linear configuration for the "reaction unit",

$$X \text{----} Y \text{----} Z$$
$$\leftarrow r_1 \rightarrow \ \leftarrow r_2 \rightarrow$$
$$\longleftarrow \quad r_3 \quad \longrightarrow$$

in which all configurations can be described by the distances r_1 and r_2, with $r_3 = r_1 + r_2$

There is no interaction between X and Y or between X and Z when X is at large distances from YZ, and so the potential energy is simply that of YZ at its equilibrium internuclear distance. When the distance between X and YZ decreases an attractive interaction is set up, and this interaction is different at different distances. A quantum mechanical calculation gives the potential energy at each distance apart, and shows that the potential energy increases as the distance r_1 decreases.

Eventually the interactions between X and Y become comparable to those between Y and Z, and this corresponds to configurations where r_1 and r_2 are comparable. The potential energy for these configurations can be calculated.

Finally the interactions between X and Y become greater than those between Y and Z. The configurations reached as Z recedes from XY and r_2 becomes greater than r_1 result in progressively decreasing potential energies. When Z is at very large distances from XY the potential energy of interaction between Z and XY is zero, and the potential energy becomes virtually that for XY at its equilibrium internuclear distance.

These configurational changes take place at constant total energy so that there is an interconversion of kinetic and potential energy resulting from the changes in configuration.

The calculations involved can be summarised in the form of a table such as:

r_1	r_2	PE
.	.	.

and the values can be plotted on a three dimensional model generating the potential energy surface.

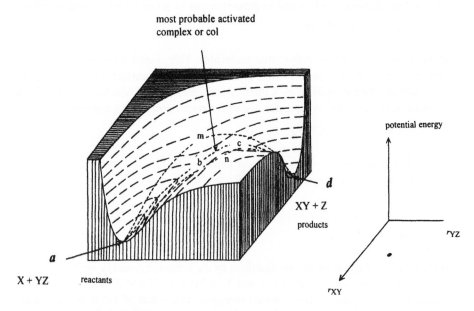

Figure 2.1   A 3-dimensional potential energy surface for a reaction between X + YZ →
XY + Z, where X, Y and Z are all atoms.

The three-dimensional surface, Figure 2.1, shows the potential energy for all possible configurations. The total energy is given as a plane through the surface, and the kinetic energy for any configuration is the vertical distance between this plane and the point representing the configuration.

2.2 Properties of the potential energy surface relevant to transition state theory: Figure 2.1

1.   The potential energy surface is unique to a given reaction. Units (X..Y..Z) with the same configuration have the same potential energy, and are at the same point on the surface, irrespective of their total energy. For instance, one point could represent a unit (X..Y..Z) which has enough energy to react, or a unit (X..Y..Z) which has not enough energy to react.

2.   On the 3-dimensional model point *a* represents reaction where X is at large distances from Y-Z. The interaction energy between X and YZ is virtually zero, and the side-view of the model shows the typical Morse potential energy curve for a diatomic molecule. The corresponding point, *d*, represents products, and gives the potential energy for molecule X-Y with Z at a large distance, and again the side view gives the typical Morse curve.

To undergo reaction the unit (X..Y..Z) must go from configuration, *a*, to configuration, *d*. Any continuous path linking point *a* to point *d*, is a physically possible route, and the route chosen by the vast majority of the units (X..Y..Z) will be the most probable one.

3.   Moving away from *a* or *d* in *all* directions results in an increase in potential energy. For every geometrically possible route between *a* and *d*, there is a point on the surface which is at a maximum potential energy. On reaction, the unit (X..Y..Z) must therefore pass through some configuration of maximum potential energy with respect to the configurations *a* and *d*. For instance paths *a*, *m*, *d* or *a*, *n*, *d* are possible routes.

4.   The Maxwell-Boltzmann distribution defines the most probable route. Low lying potential energy states are the most probable, and the probability decreases as the potential energy increases.

     Using this criterion, the two most probable states on the model are *a* the reactants, and *d* the products, while the lowest lying potential energy maximum is the most probable one. Reaction involves a series of configurational changes along the most probable route which is the sequence of lowest lying configurations which go through the lowest lying potential energy maximum, often called the minimum energy route or path. Most "units" (X..Y..Z) follow this route. However, the theory can allow for other routes through a weighting factor for each route calculated by the Maxwell-Boltzmann distribution from the potential energy for each configuration along the route.

5.   The configuration of maximum potential energy is called
     -        the *activated complex* -
     -        the *transition state* -
     -        the *critical configuration* -

     and the unit (X..Y..Z) must attain this configuration before reaction can take place. Possessing the critical energy is not sufficient, the fundamental requirement is attainment of this critical configuration.

6.   The pathway up the entrance valley, through the critical configuration of the activated complex, and into the exit valley route *a*, *b*, *c*, *d*, is called the *reaction coordinate*.

7.   The total energy is given by a plane lying horizontally through the surface. If the total energy is greater than the potential energy of the critical configuration, then the plane lies above the potential energy maximum, but it lies below if the total energy is less.

8.   The vertical distance between *a* and the critical configuration gives the necessary potential energy required for reaction. It also gives the minimum total energy for reaction since the plane through the critical configuration gives the minimum total energy which would allow the potential energy to rise to the value required for the critical configuration.

9.   Any change in the dimensions of the activated complex along the reaction coordinate leads to a decrease in potential energy. This sort of behaviour is not found in molecules, so the activated complex *cannot* be regarded as a molecule - it is merely a critical configuration. However, changes in the dimensions of the activated complex in *all* other directions result in an increase in potential energy, behaviour typical of vibration about an equilibrium internuclear distance.

10. Values of r_1 and r_2 for configurations with the same potential energy can be used to draw out a contour diagram of curves of constant potential energy, Figure 2.2.

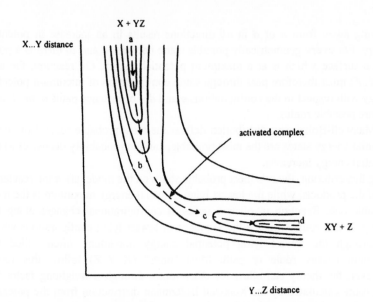

Figure 2.2   A contour diagram showing curves of constant potential energy for reaction between X + YZ → XY + Z.

The activated complex is again at the col or the saddle point.

11. Following the reaction path on the 3-dimensional surface gives a potential energy profile which shows how the potential energy changes as the reaction entity (X..Y..Z) changes configuration along the reaction coordinate, Figure 2.3.

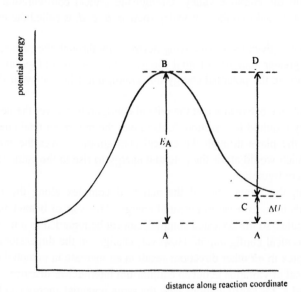

Figure 2.3 The potential energy profile for reaction X + YZ → XY + Z.

The activated complex lies at the top of the potential energy profile. The vertical distance, AB, gives the change in potential energy between reactants and activated complex, and corresponds to the critical energy required for reaction. The vertical distance, CD, gives the change in potential energy between products and activated complex while AC gives the potential energy change between reactants and products, approximately equal to ΔU^{θ} of reaction.

The potential energy profile shows very clearly the decrease in potential energy which results from any change in dimensions of the activated complex *along the reaction coordinate*.

12. An activated intermediate can be formed at any stage along the reaction pathway. It appears on the 3-dimensional surface as a well, on the contour diagram as a series of contours, and on the profile as a minimum, Figures 2.4 and 2.5.

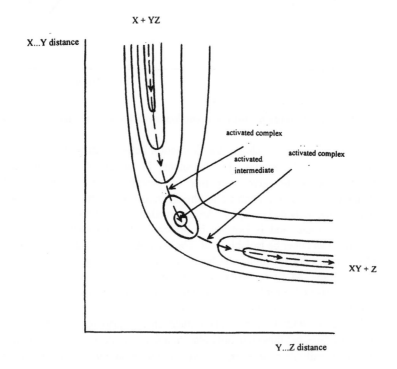

Figure 2.4   A contour diagram showing curves of constant potential energy for a reaction involving an activated intermediate.

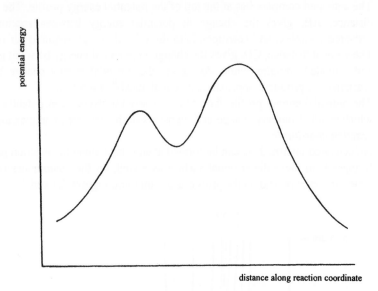

Figure 2.5 A potential energy profile for a reaction involving an activated intermediate.

    The surface also shows clearly the distinction between an activated intermediate and an activated complex. Displacements in **all** directions, even along the reaction coordinate, give **increases** in potential energy for the activated intermediate, in sharp contrast to the **decrease** along the reaction coordinate found for the activated complex. An intermediate is, therefore, a normal but high energy molecule, and is isolable in principle; the activated complex is not a molecule, can never be isolable, and is only a critical arrangement of atoms.

13. Activated or energised molecules have energy equal to, or greater than, the critical energy. They can be anywhere on the potential energy surface, but can only become activated complexes by rearranging the relative positions of the atoms in the reaction unit (X..Y..Z) until the critical configuration is reached.

    Molecules with less than the critical energy also appear on the potential energy surface, but these can only alter in a manner allowed by their total energy, and the possibilities of alteration in configuration are limited to those with potential energy equal to or less than the total energy. In particular, the critical configuration can never be reached.

14. The process of reaction is often described as "moving over the potential energy surface". This is **not** a movement in space, but only an internal movement of atoms with respect to each other. Likewise, phrases such as "surmounting a potential energy barrier" or "getting over the energy hump" and "climbing up and over the col or saddle point" are descriptive only, and the term "the rate of passage through the critical configuration" simply means a rate of change of configuration.

15. The critical configuration occupies a flat non-zero area, corresponding to a flat non-zero length along the reaction coordinate. Since the maximum is flat, the potential energy is virtually constant over that part of the surface defining the activated complex, there will be no appreciable force acting on the critical

configuration, and this results in a constant rate of change in configuration through the critical configuration. This describes a *free translation*, and passage through the critical configuration is treated as though it were a free translation over the distance defining the extent of the critical configuration along the reaction coordinate. Once the system reaches configurations with decreasing potential energies, the motion ceases to be a free translation.

16. The concentration of activated complexes is the *total* number of units (X..Y..Z) per unit volume which are in the critical configuration.

17. The rate of reaction is the total number of units (X..Y..Z) per unit volume which pass through the critical configuration per unit time *from reactant valley to product valley*.

Some units (X..Y..Z) do not pass through the critical configuration in the right direction.

(a) They may have started as products and be moving to reactants.

(b) They may be passing through the critical configuration in the right direction, but then return to the reactant valley.

(c) They may have reached the col from the reactant valley, but be deflected from the wall of the surface in such a way that the product valley cannot be reached.

(d) They may have passed over the col and then enter a well from which they emerge after a few vibrations, and return to the reactant valley.

All these systems will contribute to the concentration of activated complexes, but *only those* activated complexes coming from the reactant valley and proceeding to products will contribute to the rate. These considerations apply to any other route, though paths lying well up the sides or over the plateau make insignificant contributions to the rate compared with the minimum energy path *a, b, c, d*.

2.3 Major Assumptions of Transition State Theory

There are four basic assumptions of transition state theory and these are given below. At this stage discussion will be left until the significance of these assumptions is more obvious.

Assumption 1

When an activated complex crosses through the critical configuration it continues changing configuration in the same direction, and *cannot* return back over the col in the same direction. This assumption requires that there is a full Maxwell-Boltzmann equilibrium distribution in *reactants* and *activated complexes*.

Assumption 2

There is a full Maxwell-Boltzmann equilibrium distribution of energy within each of the *reactants*, *activated complexes*, and *products*, and, in particular, there is *equilibrium* between *reactants* and *activated complexes*. This assumption is not required if single crossings of the energy barrier are assumed, but there are difficulties when multiple crossings are allowed.

Assumption 3

Motion through the critical configuration is treated as special and as though it were a free translation, independent of the other internal movements of the activated complex.

Assumption 4
Motion on the potential energy surface can be treated classically. Ideally the movement should be described in quantum terms.

2.4 Derivation of the rate of reaction
As the unit (X..Y..Z) moves along the reaction coordinate it spends time in each configuration, and, in particular, it spends time τ in the critical configuration which extends over the length δ on the reaction coordinate. The rate at which the unit changes configuration over the length δ is

$$\frac{\delta}{\tau} = v \tag{2.1}$$

where v is often called "the rate of passage over the energy barrier", or "the velocity of getting over the energy barrier".

If $N^{\neq}$ activated complexes pass through the critical configuration in time τ, then

$$\frac{N^{\neq}}{V} \frac{1}{\tau} = c^{\neq} \frac{1}{\tau} = c^{\neq} \frac{v}{\delta} \tag{2.2}$$

pass through the critical configuration per unit volume per unit time, where $c^{\neq}$ is the concentration of activated complexes.

However, not all units (X..Y..Z) which have reached the critical configuration and become activated complexes may actually react. The transmission coefficient, κ, which is the fraction of activated complexes which actually do react, allows for this

$$\text{Rate of reaction} = \kappa c^{\neq} \frac{v}{\delta}. \tag{2.3}$$

2.4.1 Calculation of $v/\delta = 1/\tau$
Activated complexes can either spend the same time in the critical configuration, or there can be a spread in values of this time. The latter is physically more realistic, and implies a similar spread in the rate of passage through the critical configuration. The Maxwell-Boltzmann distribution gives this rate.

Rates of passage through the critical configuration from left to right have positive values from zero to infinity. Rates of motion in the opposite direction have negative values from zero to minus infinity, and they do *not* contribute to the rate of reaction because the change in configuration is from products to reactants.

The Maxwell-Boltzmann distribution gives the following expression

$$\frac{N^{\neq} \exp\left(-\dfrac{mv^2}{2kT}\right) dv}{\displaystyle\int_{-\infty}^{\infty} \exp\left(-\dfrac{mv^2}{2kT}\right) dv} \tag{2.4}$$

for the number of activated complexes which have a rate of change of configuration, v, in the range $v \to v + dv$ in **one** squared term and which pass through the critical configuration in a time τ in the range $\tau \to \tau + d\tau$. One squared term is used because only forward moving activated complexes contribute to the rate.

The number of activated complexes which pass through the critical configuration **per unit time**, and have a rate of change of configuration, v, in the range $v \to v + dv$ is

$$\frac{N^{\neq} \exp\left(-\dfrac{mv^2}{2kT}\right)dv}{\tau \displaystyle\int_{-\infty}^{\infty} \exp\left(-\dfrac{mv^2}{2kT}\right)dv}.$$
(2.5)

Substituting v/δ for $\frac{1}{\tau}$ gives

$$\frac{N^{\neq} v \exp\left(-\dfrac{mv^2}{2kT}\right)dv}{\delta \displaystyle\int_{-\infty}^{\infty} \exp\left(-\dfrac{mv^2}{2kT}\right)dv}.$$
(2.6)

The total number of activated complexes with **all** possible rates of passage, v, through the critical configuration in **one** direction from reactant valley to product valley must now be calculated. This requires summing over all positive values of v from zero to infinity. Negative values of v do not contribute to the rate.

The total number of activated complexes passing through the critical configuration from left to right per unit time per unit volume

$$= \frac{\kappa N^{\neq} \displaystyle\int_{0}^{\infty} v \exp\left(-\dfrac{mv^2}{2kT}\right)dv}{V\delta \displaystyle\int_{-\infty}^{\infty} \exp\left(-\dfrac{mv^2}{2kT}\right)dv}$$
(2.7)

$$= \kappa \frac{c^{\neq}}{\delta}\left(\frac{kT}{2\pi\mu^{\neq}}\right)^{\frac{1}{2}}$$
(2.8)

where $\mu^{\neq}$ is the reduced mass along the reaction coordinate, and is calculable from the potential energy surface.

The rate of reaction can also be expressed as

(2.9)

$$kc_X c_{YZ}$$

where c_X and c_{YZ} are the numbers of molecules per unit volume.

$$k = \kappa \frac{c^{\neq}}{c_X c_{YZ}} \frac{1}{\delta}\left(\frac{kT}{2\pi\mu^{\neq}}\right)^{\frac{1}{2}}.$$
(2.10)

2.4.2 Calculation of $c^{\neq}$ and $c^{\neq}/c_X c_{YZ}$
There is no easy way to calculate $c^{\neq}$. Equilibrium statistical mechanics can calculate

$$\left(\frac{c^{\neq}}{c_X c_{YZ}}\right)_{eq}$$

which is the quotient *at equilibrium*, but the expression appearing in transition state theory is a quotient under non-equilibrium conditions.

However, for ease of calculation and to enable statistical mechanics to be used, *equilibrium is assumed to exist between reactants and activated complexes*, assumption 1.

$$X + YZ \rightleftharpoons \text{activated complexes} \quad \rightarrow \quad XY + Z$$

with

$$K = \left(\frac{c^{\neq}}{c_X c_{YZ}}\right)_{eq}.$$

$$(2.11)$$

More complete calculations using non-equilibrium statistical mechanics give agreement with this calculation, provided $E_0 > kT$.

2.4.3 A statistical mechanical expression for the rate constant
The statistical mechanical equilibrium constant for the process

$$X + YZ \rightleftharpoons \text{activated complexes}$$

is

$$K = \frac{Q^{\neq}}{Q_X Q_{YZ}} \exp\left(-\frac{\Delta U_0}{RT}\right)$$

$$(2.12)$$

where K refers to the number of molecules per unit volume, $Q^{\neq}$ is a partition function per unit volume for the activated complex, Q_X and Q_{YZ} are partition functions per unit volume for the reactants.

ΔU_0 is ΔU at absolute zero, and is equal to E_0, the activation energy per mol of reactants converted to activated complexes. It is normally found from the potential energy surface

$$\frac{c^{\neq}}{c_X c_{YZ}} = \frac{Q^{\neq}}{Q_X Q_{YZ}} \exp\left(-\frac{E_0}{RT}\right)$$

$$(2.13)$$

$$k = \kappa \frac{1}{\delta} \frac{Q^{\neq}}{Q_X Q_{YZ}} \left(\frac{kT}{2\pi\mu^{\neq}}\right)^{\frac{1}{2}} \exp\left(-\frac{E_0}{RT}\right)$$

$$(2.14)$$

which is the basic equation of transition state theory in partition function form.

2.4.4 Statistical mechanical quantities used in transition state theory
Transition state theory uses the molecular partition function per unit volume Q. This is related to the molecular partition function f, defined as a sum over all quantum states of the molecule in volume V.

$$f = \sum_i \exp\left(-\frac{\varepsilon_i}{kT}\right)$$ (2.15)

where ε_i is the energy of the i'th state, and the sum is over all quantum states.

An analogous description can be given as a sum over all energies

$$f = \sum_j g_j \exp\left(-\frac{\varepsilon_j}{kT}\right)$$ (2.16)

where ε_j is the energy of the j'th state, g_j is the number of states with energy ε_j, and the sum is over all energies ε_j.

The energy of a state is made up of translational, rotational, vibrational and electronic energy, and to a first approximation these are assumed to be independent of each other. The total energy of the i'th quantum state can then be written as a sum

$$\varepsilon_i = \varepsilon_{trans} + \varepsilon_{rot} + \varepsilon_{vib} + \varepsilon_{el}$$ (2.17)

which implies that

$$f = f_{trans}\, f_{rot}\, f_{vib}\, f_{el}.$$ (2.18)

Table 2.1 gives the standard forms for each individual partition function.

Table 2.1 Statistical mechanical quantities relevant to kinetics

Type of partition function	Defining relation	Standard expression	Quantities involved other than universal constants and temperature
translation	$f_{trans} = \sum_i \exp\left(-\dfrac{\varepsilon_{trans}}{kT}\right)$ sum over all quantum states i	$\dfrac{(2\pi mkT)^{3/2} V}{h}$	m = mass of molecule
rotation: linear molecule	$f_{rot} = \sum_i \exp\left(-\dfrac{\varepsilon_{rot}}{kT}\right)$ sum over all quantum states i	$\dfrac{8\pi^2 IkT}{\sigma h^2}$	σ = rotational symmetry number I = moment of inertia found from spectroscopic data
rotation: non-linear molecule	$f_{rot} = \sum_i \exp\left(-\dfrac{\varepsilon_{rot}}{kT}\right)$ sum over all quantum states i	$\dfrac{8\pi^2 (8\pi^3 I_A I_B I_C)^{1/2} (kT)^{3/2}}{\sigma h^3}$	σ = rotational symmetry number I_A, I_B, I_C are the three moments of inertia about three axes found from spectroscopic data

rotation: internal

$$f_{int\,rot} = \sum_i \exp\left(-\frac{\varepsilon_{rot}}{kT}\right)$$

sum over all quantum states i

$$\frac{(8\pi^3 I' kT)^{1/2}}{\sigma' h}$$

if there is free rotation

σ' = symmetry number for the internal rotation. I' is the moment of inertia for the internal rotation found from spectroscopic data

vibration for a single mode

$$f_{vib} = \sum_i \exp\left(-\frac{\varepsilon_{vib}}{kT}\right)$$

sum over all quantum states i

$$\frac{1}{1-\exp\left(\frac{-h\nu}{kT}\right)}$$

ν is the fundamental vibration frequency for a given vibrational mode

vibration for all modes of a molecule

$$f_{vib} = \sum_i \exp\left(-\frac{\varepsilon_{vib}}{kT}\right)$$

sum over all quantum states i

$$\prod_j \left[\frac{1}{1-\exp\left(\frac{-h\nu}{kT}\right)}\right]$$

all modes j

ν_j are the various fundamental vibration frequencies for the vibrational modes of the molecule

Number of vibrational modes for a linear molecule = 3N-5
Number of vibrational modes for a non-linear molecule = 3N-6
Number of vibrational modes for a non-linear molecule with x internal rotations = 3N-x-6

electronic
ground state

$$f_{el} = \sum_i \exp\left(-\frac{\varepsilon_{el}}{kT}\right)$$

sum over all quantum
states i

g_0 if only the lowest energy
level needs to be
considered

the electronic energy levels must be known
explicitly and are found from a
spectroscopic analysis. Often $f_{el} = 1$ for the
ground electronic state

$$g_0 + g_1 \exp\left(\frac{-\varepsilon_1}{kT}\right) +$$

$$g_2 \exp\left(\frac{-\varepsilon_2}{kT}\right) + \cdots \text{ where } 1, 2 \ldots$$

are the degeneracies of the first,
second ... excited levels

h = Planck's constant V = volume

k = Boltzmann's constant T = absolute temperature

2.4.5 Relation between the molecular partition function f, and the molecular partition function per unit volume

The total molecular partition function, f, is

$$f = f_{trans}\, f_{rot}\, f_{vib}\, f_{el} \tag{2.19}$$

and the translational partition function depends on the volume V

$$f_{trans} = \frac{(2\pi mkT)^{\frac{3}{2}}}{h^3} V \tag{2.20}$$

$$\therefore \frac{f}{V} = \frac{(2\pi mkT)^{\frac{3}{2}}}{h^3} f_{rot}\, f_{vib}\, f_{el} \tag{2.21}$$

where f/V is the molecular partition function per unit volume, Q, which appears in transition state theory.

$$\therefore \quad Q = \frac{(2\pi mkT)^{\frac{3}{2}}}{h^3} f_{rot}\, f_{vib}\, f_{el} \tag{2.22}$$

$$= q\, f_{int} \tag{2.23}$$

where f_{int} is the molecular partition for internal motion and

$$q = \frac{(2\pi mkT)^{\frac{3}{2}}}{h^3}. \tag{2.24}$$

2.4.6 Contributions to the partition functions for polyatomic molecules and radicals

Since most reactions occur in the ground electronic state, the discussion will be limited to contributions from translation, rotation and vibration.

The total number of degrees of freedom for a polyatomic molecule containing N atoms is $3N$. Of these, three are taken up by translational motion, two by rotation when the molecule is linear, and three by rotation when non-linear. Internal rotations, or other types of motion contributing to the energy must be included before the number of vibrational degrees of freedom allowable can be found. For instance, a molecule could have three translations, three rotations and two internal rotations, and the number of vibrations then is

$$3N - 3(\text{trans}) - 3(\text{rot}) - 2(\text{int rot}) = 3N\text{-}8$$

If there are no internal rotations the number of vibrations which are possible are $3N$-5 and $3N$-6 for linear and non-linear molecules respectively, giving $3N$-5 and $3N$-6 terms for the vibrational partition function.

2.4.7 Contributions to the partition function for the activated complex

The activated complex has the unique feature of a free internal translation along the reaction coordinate, and this must be included when the number of vibrational terms is being considered. Of the total of $3N$ terms

(1)   three are required for translation
(2)   two or three are required for rotation
(3)   one is required for the free translation over the length, δ, of the critical configuration.

The inclusion of the free translation when working out the terms other than vibrational terms, means that an *activated complex has always one degree of vibrational freedom less* than a molecule containing the same number of atoms, one of the vibrational degrees of freedom having degenerated into the free translation along the reaction coordinate.

The molecular partition function for this free translation along a length δ is

$$f_{\text{free trans}} = \frac{(2\pi\mu^{\ddagger} kT)^{\frac{1}{2}}}{h} \delta.$$

(2.25)

2.4.8 Simplification of the statistical mechanical transition state theory expression

At this stage it is sensible to generalise to reaction between two molecules X and YZ, remembering that this requires an n dimensional surface.

$$k = \kappa \frac{1}{\delta} \frac{Q^{\ddagger}}{Q_X Q_{YZ}} \left(\frac{kT}{2\pi\mu^{\ddagger}} \right)^{\frac{1}{2}} \exp\left(-\frac{E_0}{RT} \right).$$

(2.26)

The terms δ and $\mu^{\ddagger}$ can be rigorously defined in terms of the potential energy surface, but use of the statistical mechanical formulation of the total molecular partition function for the activated complex enables these to be cancelled out without quantifying either δ or $\mu^{\ddagger}$.

Equation 2.22 gives the molecular partition function for any molecule

$$Q = q\, f_{\text{rot}}\, f_{\text{vib}}\, f_{\text{el}}$$

(2.27)

$$= q\, f_{\text{int}}$$

where

$$q = \frac{(2\pi m kT)^{\frac{3}{2}}}{h^3}$$

(2.28)

The corresponding expression for the activated complex is

$$Q^{\ddagger} = q^{\ddagger}\, f_{\text{int}}^{\ddagger}.$$

(2.29)

The internal free translation, which is the special feature of an activated complex, can be factorised out from f_{int} giving

$$Q^{\ddagger} = q^{\ddagger} f^{\ddagger}_{free\ trans}\ f^{\ddagger *}_{int} \qquad (2.30)$$

where f^{*}_{int} is the molecular partition function *left over after the free translation has been separated out*.

$$f^{\ddagger}_{free\ trans} = \frac{\left(2\pi\mu^{\ddagger}kT\right)^{\frac{1}{2}}\delta}{h} \qquad (2.31)$$

hence

$$Q^{\ddagger} = q^{\ddagger} f^{\ddagger *}_{int} \frac{\left(2\pi\mu^{\ddagger}kT\right)^{\frac{1}{2}}\delta}{h}$$

$$= Q^{\ddagger *} \frac{\left(2\pi\mu^{\ddagger}kT\right)^{\frac{1}{2}}\delta}{h} \qquad (2.32)$$

where $Q^{\ddagger *}$ is the molecular partition function for the activated complex where one term for free translation has been removed, and where $Q^{\ddagger *}$ has, in consequence, one vibrational term fewer than has a polyatomic molecule with the same numbers of atoms.

This expression can be substituted for $Q^{\ddagger}$ in Equation 2.14 giving

$$k = \frac{\kappa}{\delta} \frac{Q^{\ddagger *}}{Q_X Q_{YZ}} \frac{\left(2\pi\mu^{\ddagger}kT\right)^{\frac{1}{2}}\delta}{h} \left(\frac{kT}{2\pi\mu^{\ddagger}}\right)^{\frac{1}{2}} \exp\left(-\frac{E_0}{RT}\right) \qquad (2.33)$$

$$= \kappa \frac{kT}{h} \frac{Q^{\ddagger *}}{Q_X Q_{YZ}} \exp\left(-\frac{E_0}{RT}\right) \qquad (2.34)$$

where the awkward terms $\mu^{\ddagger}$ and δ have cancelled out.

The final step requires $Q^{\ddagger *}$ for the activated complex, and the Q's for the reactants to be calculated.

2.4.9 Formulation of Q for reactants

Table 2.1 gives the expressions for the molecular partition functions. The quantities involved can be calculated from spectroscopic data; moments of inertia appear in rotational partition functions, and fundamental vibrational frequencies of each mode of vibration are required for the vibrational partition functions. The number of terms involved is given in Table 2.2. This is for reactions occurring in the ground electronic state. If excited states are involved the contribution to Q from the electronic motion must be considered before the number of vibrational terms are found. But this would require motion over more than one potential energy surface.

Table 2.2 Contributions to the total molecular partition function per unit volume

(a) <u>Reactant molecules</u>

Type of molecule	Total number of degrees of freedom available	Total number of degrees of freedom available to translation	Total number of degrees of freedom available to rotation	Total number of degrees of freedom available to internal rotation	Total number of degrees of freedom left for vibration
atom	$3N$	3	-	-	-
linear molecule	$3N$	3	2	x	$3N - 5 - x$
non-linear molecule	$3N$	3	3	x	$3N - 6 - x$

(b) <u>Activated Complex</u>

Type of activated complex	Total number of degrees of freedom	Total number of degrees of freedom available to translation	Total number of degrees of freedom available to rotation	Total number of degrees of freedom available to internal rotation	Number of degrees of freedom allocated to the free translation	Total number of degrees of freedom left for vibration
linear	$3N$	3	2	x	1	$3N - 6 - x$
non-linear	$3N$	3	3	x	1	$3N - 7 - x$

2.4.10 Formulation of $Q^{\neq*}$ for the activated complex

The same formal expressions, Table 2.1 are required for translational, rotational and vibrational partition functions for the activated complex. Table 2.2 gives the number of vibrational terms open to the activated complex when one vibrational term is lost and the free translation gained.

$Q^{\neq*}$ can be found provided the potential energy surface is known. The position of the critical configuration gives the interatomic distances from which the moments of inertia can be calculated, while the barrier curvature and the interatomic distances give the frequencies of vibration for the normal modes of vibration. Unfortunately the region of the critical configuration is the most difficult part of the potential energy surface to detail rigorously, and this can be the limiting factor in the accuracy of a purely theoretical calculation of $Q^{\neq*}$.

Modifications in the expressions for the partition functions, are sometimes necessary for both reactants and activated complexes. For instance, the expressions for translation and rotation assume small spacings in the energy levels. This assumption is always valid for translation, but is only valid for rotations at high enough temperatures. The expression for vibrations assumes harmonic oscillators, valid only at low temperatures.

When the surface cannot be obtained, or cannot be obtained sufficiently accurately, it is possible to estimate $Q^{\neq*}$ by analogy with a stable molecule of similar structure.

Table 2.3 gives actual values of the quotient

$$\kappa \frac{kT}{h} \frac{Q^{\neq*}}{Q_X Q_{YZ}}.$$

This shows the remarkably good predictions made by transition state theory, and can be compared with the much poorer predictive value of collision theory as illustrated in Table 2.4.

Table 2.3   Comparison of experimental A factors with pre-exponential factors calculated from transition state theory

Reaction			$\log A_{10}$ observed $A/\text{mol}^{-1} \text{dm}^3 \text{s}^{-1}$	$\log A_{10}$ calculated $A/\text{mol}^{-1} \text{dm}^3 \text{s}^{-1}$
$2ClO\cdot$	$\rightarrow$	$Cl_2 + O_2$	7.7	7.0
$F_2 + ClO_2$	$\rightarrow$	$FClO_2 + F\cdot$	7.5	7.9
$NO_2 + CO$	$\rightarrow$	$NO + CO_2$	10.1	9.8
$2NO_2$	$\rightarrow$	$2NO + O_2$	9.3	9.7
$2NOCl$	$\rightarrow$	$2NO + Cl_2$	10.0	8.6
$NO + Cl_2$	$\rightarrow$	$NOCl + Cl\cdot$	9.6	9.1
$NO_2 + F_2$	$\rightarrow$	$NO_2F + F\cdot$	9.2	8.1
$H\cdot + H_2$	$\rightarrow$	$H_2 + H\cdot$	10.7	10.9
$H\cdot + CH_4$	$\rightarrow$	$CH_3\cdot + H_2$	10.0	10.3
$H\cdot + C_2H_6$	$\rightarrow$	$C_2H_5\cdot + H_2$	9.5	10.0
$CH_3\cdot + C_2H_6$	$\rightarrow$	$CH_4 + C_2H_5\cdot$	9.3	8.0
$Br_2 + H\cdot$	$\rightarrow$	$HBr + Br\cdot$	10.5	11.0

Table 2.4   Bimolecular reactions: comparison of experimental A factors with collision numbers Z calculated from collision theory

Reaction			$\log_{10}A$ $A/\text{mol}^{-1}\ \text{dm}^3\ \text{s}^{-1}$	$\log_{10}Z$ $Z/\text{mol}^{-1}\ \text{dm}^3\ \text{s}^{-1}$
$NO_2 + F_2$	$\rightarrow$	$NO_2F + F\cdot$	9.2	10.8
$2NO_2$	$\rightarrow$	$2NO + O_2$	9.3	10.6
$2NOCl$	$\rightarrow$	$2NO + Cl_2\cdot$	10.0	10.8
$2ClO\cdot$	$\rightarrow$	$Cl_2 + O_2$	7.8	10.4
$F_2 + ClO_2$	$\rightarrow$	$FClO_2 + F\cdot$	7.5	10.7

2.5 Approximate calculations

Accurate calculations as given in Table 2.3 rely on spectroscopic analysis of the reactants, and on the potential energy surface for specific reactions, and by assuming that the experimental activation energy E_A can be compared with the activation energy, E_0, calculated from the potential energy surface, then the above quotient can be compared with the experimental A factor.

Qualitative comparisons can, however, be made using rough order of magnitude partition functions for translation, rotation and vibration. This approximation is acceptable because partition functions do not vary much with mass, moment of inertia or vibrational frequency.

In transition state theory, the quotient

$$\kappa \frac{kT}{h} \frac{Q^{\ddagger*}}{Q_X Q_{YZ}}$$

can be compared with

1. The A factor in the Arrhenius equation

$$k = A \exp\left(-\frac{E_A}{RT}\right). \tag{2.35}$$

2. The collision number Z in simple collision theory

$$k = pZ \exp\left(-\frac{E_A}{RT}\right). \tag{2.36}$$

The molecular partition function has been defined as

$$Q = q\, f_{rot}\, f_{vib} \tag{2.37}$$

where q, f_{rot}, f_{vib} contain all the contributions from translation, rotation and vibration. Each of these can in turn be expressed as a product of terms each of which refers to single degrees of freedom for the motion involved,

       q is a product of three factors, q_t

       f_{rot} is a product of two or three factors, f_r

       f_{vib} is a product of factors, f_v, with one for each mode of vibration

and each of q_t, f_r and f_v is not expected to vary much from molecule to molecule.
Typical values for each q_t, f_r and f_v are

$$q_t \quad = \quad 10^9 \text{ to } 10^{10} \text{ dm}^{-1}$$
$$f_r \quad = \quad 10 \text{ to } 10^2$$
$$f_v \quad = \quad 1 \text{ to } 10$$

Table 2.4 shows that, unlike Z, the experimental A factor varies considerably from reaction to reaction. These discrepancies are typical, and the conclusion is that simple collision theory is inadequate, and any agreement found is fortuitous. This is not surprising since simple collision theory assumes the reactants to be hard spheres which is physically unrealistic. Also, the internal structures of the molecules are ignored.

Transition state theory considers the internal structures of the reactants and activated complex explicitly, either in the accurate calculations given in Table 2.3, or in the approximate calculations given for several reaction types, Table 2.5.

Table 2.5 Calculation of pre-exponential factors from transition state theory using approximate values for the partition functions; κ is the transmission coefficient

Reaction	Partition Functions	$\dfrac{Q^{\neq *}}{Q_X Q_{YZ}}$	$\dfrac{\kappa kT}{h} \dfrac{Q^{\neq *}}{Q_X Q_{YZ}}$	Value/dm³ sec⁻¹
Two atoms A + B → A.C.	A q_t^3 B q_t^3 A.C. $q_t^3 f_r^2$	$\dfrac{q_t^3 f_r^2}{q_t^3 q_t^3}$	$\dfrac{\kappa kT}{h} \dfrac{f_r^2}{q_t^3}$	$10^{-12} - 10^{-13}$
Atom + linear molecule → linear activated complex A + M → A.C.	A q_t^3 M $q_t^3 f_r^2 f_v^{3N-5}$ A.C. $q_t^3 f_r^2 f_v^{3N-6}$ where $N' = N + 1$ ∴ A.C. $q_t^3 f_r^2 f_v^{3N-3}$	$\dfrac{q_t^3 f_r^2 f_v^{3N-3}}{q_t^3 f_r^2 f_v^{3N-5} q_t^3}$	$\dfrac{\kappa kT}{h} \dfrac{f_v^2}{q_t^3}$	$10^{-15} - 10^{-14}$
Atom + non-linear molecule → non-linear activated complex A + M → A.C.	A q_t^3 M $q_t^3 f_r^3 f_v^{3N-6}$ A.C. $q_t^3 f_r^3 f_v^{3N-7}$ where $N' = N + 1$ ∴ A.C. $q_t^3 f_r^3 f_v^{3N-4}$	$\dfrac{q_t^3 f_r^3 f_v^{3N-4}}{q_t^3 q_t^3 f_r^3 f_v^{3N-6}}$	$\dfrac{\kappa kT}{h} \dfrac{f_v^2}{q_t^3}$	$10^{-15} - 10^{-14}$

Reaction	Partition functions		Value/s⁻¹
Two linear molecules → non-linear activated complex	M $q_t^3 f_r^2 f_v^{3N-5}$ N $q_t^3 f_r^2 f_v^{3N'-5}$ A.C. $q_t^3 f_r^3 f_v^{3N''-7}$ where $N'' = N + N'$ ∴ A.C. $q_t^3 f_r^3 f_v^{3N+3N'-7}$	$\dfrac{q_t^3 f_r^3 f_v^{3N''-7}}{q_t^3 f_r^2 q_t^3 f_r^2 f_v^{\,3N+3N'-10}}$ $\qquad \dfrac{\kappa kT}{h}\,\dfrac{f_v^3}{q_t^3 f_r}$	$10^{-16} - 10^{-15}$
M + N → A.C.			
Two non-linear molecules → non-linear activated complex	M $q_t^3 f_r^3 f_v^{3N-6}$ N $q_t^3 f_r^3 f_v^{3N'-6}$ A.C. $q_t^3 f_r^3 f_v^{3N''-7}$ where $N'' = N + N'$ ∴ A.C. $q_t^3 f_r^3 f_v^{3N+3N'-7}$	$\dfrac{q_t^3 f_r^3 f_v^{\,3N+3N'-7}}{q_t^3 f_r^3 q_t^3 f_r^3 f_v^{\,3N+3N'-12}}$ $\qquad \dfrac{\kappa kT}{h}\,\dfrac{f_v^5}{q_t^3 f_r^3}$	$10^{-18} - 10^{-17}$
M + N → A.C.			Value/s⁻¹
linear molecule → non-linear activated complex	M $q_t^3 f_r^2 f_v^{3N-5}$ A.C. $q_t^3 f_r^3 f_v^{3N-7}$	$\dfrac{q_t^3 f_r^3 f_v^{\,3N-7}}{q_t^3 f_r^2 f_v^{\,3N-5}}$ $\qquad \dfrac{\kappa kT}{h}\,\dfrac{f_r}{f_v^2}$	$10^{14} - 10^{17}$
M → A.C.			
non-linear → non-linear activated complex	M $q_t^3 f_r^3 f_v^{3N-6}$ A.C. $q_t^3 f_v^{3N-7} f_r^3$	$\dfrac{q_t^3 f_r^3 f_v^{\,3N-7}}{q_t^3 f_r^3 f_v^{\,3N-6}}$ $\qquad \dfrac{\kappa kT}{h}\,\dfrac{1}{f_v}$	$10^{13} - 10^{14}$
M → A.C.			

OTHER EXAMPLES BEING CALCULATED SIMILARLY

2.5.1 Reaction of two atoms, A and B, to give a linear diatomic activated complex

$$A + B \rightleftharpoons AB^* \rightarrow \quad \text{product}$$

where

$$Q_A = q_t^3$$
$$Q_B = q_t^3 \qquad \text{there are no rotations or vibrations for an atom}$$

$$Q^{**} = q_t^3 f_r^2 \qquad \begin{array}{l}\text{a diatomic species has two rotations, and the}\\ \text{activated complex has one vibration less than the}\\ \text{corresponding diatomic molecule.}\end{array}$$

giving

$$\kappa \frac{kT}{h} \frac{Q^{**}}{Q_A Q_B} \tag{2.38}$$

which is equal to

$$\kappa \frac{kT}{h} \frac{f_r^2}{q_t^3}. \tag{2.39}$$

Typical values of the quotient lie between 6×10^{11} and 6×10^{10} mol^{-1} dm^3 s^{-1} which compares well with the simple collision theory value of around 10^{11} mol^{-1} dm^3 s^{-1}, even though the two theories are conceptually far apart.

In this reaction, three degrees of freedom for translational motion are being replaced by two degrees of freedom for rotational motion. Collision theory requires consideration of translational motion, but does not recognise the relevance of rotational motion, which will be shown later to be the dominant feature in determining the magnitude of the A factor.

2.5.2 A more general reaction: two non-linear molecules, A and B, containing N_A and N_B atoms respectively, giving a non-linear activated complex which contains $N_A + N_B$ atoms

$$Q_A = q_t^3 f_r^3 f_v^{3N_A - 6}$$
$$Q_B = q_t^3 f_r^3 f_v^{3N_B - 6}$$
$$Q^{**} = q_t^3 f_r^3 f_v^{3(N_A + N_B) - 7}$$

(Q^{**} has one vibration fewer than the polyatomic molecule containing $3(N_A + N_B)$ atoms) giving

$$\kappa \frac{kT}{h} \frac{Q^{**}}{Q_A Q_B} = \kappa \frac{kT}{h} \frac{f_v^5}{q_t^3 f_r^3}. \tag{2.40}$$

Typical values of the quotient lie in the range 6×10^6 — 6×10^5 mol^{-1} dm^3 s^{-1} which is considerably different from the collision theory value of around 10^{11} mol^{-1} dm^3 s^{-1}.

In this reaction three translational and three rotational degrees of freedom are lost and are replaced by five vibrational degrees of freedom.

2.5.3   Approximate transition state theory pre-exponential factors for typical reaction types

Tables 2.3 and 2.5 shows that the transition state theory pre-exponential factor varies over several powers of ten, compared to the much smaller range 10^{10} to 10^{12} mol^{-1} dm^3 s^{-1} given by collision theory, Table 2.4.

Collision theory can be brought into line with experiment via the p factor calculated from

$$A = pZ$$

where p is a measure of the deviation of A from a simple collision number for two hard spheres.

As shown earlier section (2.5.1) the transition state theory analogue of the collision number is

$$\kappa \frac{kT}{h} \frac{f_r^2}{q_t^3}$$

and so values of the quotient can be compared directly with experimental values of

$$\frac{\kappa kT}{h} \frac{Q^{\neq *}}{Q_X Q_{YZ}} \bigg/ \frac{\kappa kT}{h} \frac{f_r^2}{q_t^3}$$

can be compared directly with experimental valies of "p". Values of the quotient are given in Table 2.6 and show that in this approximate analysis transition state theory gives a satisfactory interpretation of the experimental "p" factors. As the complexity of the reactants increases the value of the quotient becomes less, as does p

Table 2.6 Values of log A calculated from transition state theory showing the dependence of $\dfrac{\kappa kT}{h} \cdot \dfrac{Q^{\neq}}{Q_{\text{reactants}}}$ on molecular complexity, and comparison with experiment. Value of $\log_{10} Z$ calculated from collision theory is 11.2 (Z is in units $\text{mol}^{-1}\,\text{dm}^3\,\text{s}^{-1}$)

Reaction	$\log_{10}\left(\dfrac{A_{\text{calc}}}{\text{dm}^3\,\text{mol}^{-1}\,\text{s}^{-1}}\right)$	$\log_{10}\left(\dfrac{A_{\text{obs}}}{\text{dm}^3\,\text{mol}^{-1}\,\text{s}^{-1}}\right)$	p_{calc}	p_{obs} (average)
Br· + H$_2$ → HBr + Br·	11.1	10.4 to 11.4	8×10^{-1}	5×10^{-1}
H· + H$_2$ → H$_2$ + H·	10.7	10.7 to 11.1	3×10^{-1}	3×10^{-1}
H· + CH$_4$ → H$_2$ + CH$_3$·	10.3	8.5 to 11.5	1.3×10^{-1}	6×10^{-2}
H· + C$_2$H$_6$ → H$_2$ + C$_2$H$_5$·	10.1	9.5 to 11.1	8×10^{-2}	13×10^{-2}
CH$_3$· + H$_2$ → CH$_4$ + H·	9.2	8.7 to 9.5	1.0×10^{-2}	8×10^{-3}
CH$_3$· + C$_2$H$_6$ → CH$_4$ + C$_2$H$_5$·	8.0	7.8 to 9.3	6×10^{-4}	2×10^{-3}
CH$_3$· + CH$_3$COCH$_3$ → CH$_4$ + ·CH$_2$COCH$_3$	8.0	8.4 to 8.8	6×10^{-4}	3×10^{-3}
F$_2$ + ClO$_2$ → FClO$_2$ + F·	7.9	7.1 to 7.6	5×10^{-4}	2×10^{-4}
CH$_3$· + HC(CH$_3$)$_3$ → CH$_4$ + ·C(CH$_3$)$_3$	6.8	7.0 to 8.5	4×10^{-5}	3.5×10^{-4}

The observed A factors can therefore be interpreted in terms of the replacement of the internal vibrations in the reactants by others in the activated complex, and the loss of translational terms, q_t, and rotational terms, f_r. Since the magnitude of f_r can lie anywhere in the range 10 to 100 whereas that for f_v lies between 1 and 10, the dominant effect is loss of rotations. There is also loss of translational degrees of freedom, but as this is the same throughout any series of reactions, loss of translational degrees of freedom will not be responsible for the trend in the quotient,

$$\kappa \frac{kT}{h} \frac{Q^{\neq *}}{Q_X Q_{YZ}}$$

and in the "calculated p" with complexity of reaction type. Table 2.6 shows that the "p" factor becomes smaller as the complexity increases.

The translational and rotational motion is motion in space, and the internal structure of the reactants or activated complex will not have a *major* effect on the actual magnitudes of q_t or f_r. The approximation that q_t or f_r have the same values for all reaction types seems to be reasonable. However, vibrational motion is internal and is characteristic of the molecule in question. Furthermore "bond breaking and bond making" could have a significant effect on the magnitude of the fundamental vibrational frequencies for the activated complex, leading to different values for the various f_v terms.

To carry these arguments any further, it would be necessary to move to the exact calculation of all the partition functions involved, such as is summarised in Table 2.3.

2.5.4 Temperature dependence of partition functions and A factors
Table 2.1 gives the expressions for the partition functions for translation, rotation and vibration, from which the temperature dependence can be found.
 (i) the translational partition function for each degree of freedom depends on $T^{1/2}$
 (ii) the rotational partition function for each degree of freedom depends on $T^{1/2}$
 (iii) the vibrational partition function for each degree of freedom is independent of temperature at low temperatures but depends on T^1 at high temperatures.

This can be used to calculate the temperature dependence of the A factors for specific reaction types. For instance, for reaction of two non-linear molecules A and B to give a non-linear activated complex, Section 2.5.2, the pre-exponential factor is given by

$$\kappa \frac{kT}{h} \frac{f_v^5}{q_t^3 f_r^3}$$

with the temperature dependence at low temperatures being

$$\frac{T}{T^{3/2} T^{3/2}} = T^{-2}$$

and at high temperatures being

$$\frac{T\,T^5}{T^{\frac{3}{2}}\,T^{\frac{3}{2}}} = T^3.$$

Other reaction types can be worked out similarly. When there is a dependence of the pre-exponential factor on temperature, it would be expected to correspond to a non-linear Arrhenius plot. However, the range of temperatures studied, and the accuracy with which the experimental determinations are made are such that the temperature dependence is slight and the curvature of the Arrhenius plot is not easily picked up.

2.6 Thermodynamic formulations of transition state theory

There are some situations, such as reactions in solution, where the partition function form is not immediately useful, whereas a thermodynamic formulation is more immediately applicable.

The equilibrium constant for formation of the activated complex from reactants is

$$K^{\neq*} = \frac{Q^{\neq*}}{Q_X Q_{YZ}} \exp\left(-\frac{E_0}{RT}\right). \tag{2.42}$$

This can be substituted into Equation 2.34 giving

$$k = \frac{\kappa k T}{h} K^{\neq*}. \tag{2.43}$$

By applying standard thermodynamic functions to the equilibrium, an expression for the rate constant in terms of $\Delta G^{\neq*}$, $\Delta H^{\neq*}$, $\Delta S^{\neq*}$ can be given

$$\Delta G^{\neq*} = -RT \ln K^{\neq*} \tag{2.44}$$

$$\Delta G^{\neq*} = \Delta H^{\neq*} - T\Delta S^{\neq*} \tag{2.45}$$

$$k = \frac{\kappa k T}{h} \exp\left(-\frac{\Delta G^{\neq*}}{RT}\right) \tag{2.46}$$

$$k = \kappa \frac{kT}{h} \exp\left(-\frac{\Delta H^{\neq*}}{RT}\right) \exp\left(\frac{\Delta S^{\neq*}}{R}\right) \tag{2.47}$$

where $\Delta G^{\neq*}$ is the free energy of activation
               $\Delta H^{\neq*}$ is the enthalpy of activation
               $\Delta S^{\neq*}$ is the entropy of activation.
The dependence of the rate constant on pressure allows $\Delta V^{\neq*}$, the volume change on activation to be found.

$K^{\neq*}$, $\Delta G^{\neq*}$, $\Delta H^{\neq*}$, $\Delta S^{\neq*}$, $\Delta V^{\neq*}$ all have *one term missing* in comparison with K, ΔG^θ, ΔH^θ, ΔS^θ, ΔV^θ describing equilibrium in a chemical reaction.

The thermodynamic functions for activation are standard values, with the standard state normally taken to be unit concentration.

2.6.1 Determination of thermodynamic functions for activation

Detailed calculations using partition functions calculated from spectroscopic data for reactants and the potential energy surface for the activated complex allow the quotient

$$\kappa \frac{kT}{h} \frac{Q^{\neq *}}{Q_X Q_{YZ}}$$

to be calculated, and from this $K^{\neq *}$, $\Delta G^{\neq *}$, $\Delta H^{\neq *}$, $\Delta S^{\neq *}$ and $\Delta V^{\neq *}$ can be found.

Experimental data can be manipulated to give experimental values to compare direct with the theoretical ones.

$$k = \kappa \frac{kT}{h} \exp\left(-\frac{\Delta H^{\neq *}}{RT}\right) \exp\left(\frac{\Delta S^{\neq *}}{R}\right),$$

(2.48)

$$\ln \frac{k}{T} = \ln \kappa \frac{k}{h} + \frac{\Delta S^{\neq *}}{R} - \frac{\Delta H^{\neq *}}{RT}$$

(2.49)

and so a plot of $\ln k/T$ against $1/T$ has

$$\text{slope} = -\frac{\Delta H^{\neq *}}{R}$$

$$\text{intercept} = \ln \kappa \frac{k}{h} + \frac{\Delta S^{\neq *}}{R}$$

$$\Delta G^{\neq *} = -RT \ln K^{\neq *}$$

and

$$\Delta G^{\neq *} = \Delta H^{\neq *} - T\Delta S^{\neq *}.$$

(2.50)

Similarly the effect of pressure on the rate constant gives $\Delta V^{\neq *}$ to compare direct with the theoretical value.

$$k = \kappa \frac{kT}{h} K^{\neq *}$$

(2.51)

$$\ln k = \ln \kappa \frac{kT}{h} - \frac{\Delta G^{\neq *}}{RT}$$

(2.52)

$$\frac{d \ln k}{dp} = -\frac{1}{RT} \left(\frac{\partial \Delta G^{\neq*}}{\partial p} \right)_T \tag{2.53}$$

$$= -\frac{\Delta V^{\neq*}}{RT}. \tag{2.54}$$

And so a plot of $\ln k$ against p has

$$\text{slope} = -\frac{\Delta V^{\neq*}}{RT}.$$

An alternative procedure relates the observed activation energy E_A, with $\Delta H^{\neq*}$, rather than obtaining $\Delta H^{\neq*}$ from a plot of $\ln k/T$ against $1/T$.

Since $K^{\neq*}$ is a standard value with the standard state being unit concentration, then

$$\frac{d \ln K^{\neq*}}{dT} = \frac{\Delta U^{\neq*}}{RT^2}. \tag{2.55}$$

But

$$k = \kappa \frac{kT}{h} K^{\neq*} \tag{2.56}$$

and

$$\frac{d \ln k}{dT} = \frac{1}{T} + \frac{d \ln K^{\neq*}}{dT} \tag{2.57}$$

$$\frac{E_A}{RT^2} = \frac{1}{T} + \frac{\Delta U^{\neq*}}{RT^2}. \tag{2.58}$$

$$\Delta U^{\neq*} = \Delta H^{\neq*} - p\Delta V^{\neq*} \tag{2.59}$$

$$E_A = \Delta H^{\neq*} - p\Delta V^{\neq*} + RT. \tag{2.60}$$

The explicit relation of E_A with $\Delta H^{\neq*}$ depends on the type of reaction involved, since this determines the value of the $p\Delta V^{\neq*}$. For gas phase reactions

$$p\Delta V^{\neq*} = \Delta n^{\neq} RT \tag{2.61}$$

where $\Delta n^{\neq*}$ is the change in the number of molecules in going from reactants to activated complexes.

For unimolecular reactions $\Delta n^{\neq} = 0$

$$p\Delta V^{\neq*} = 0 \tag{2.62a}$$

so that

$$E_A = \Delta H^{\neq*} + RT \tag{2.63a}$$

and

$$k = e\kappa \frac{kT}{h} \exp\left(-\frac{E_A}{RT}\right) \exp\left(\frac{\Delta S^{\neq*}}{R}\right). \tag{2.64}$$

For bimolecular reactions $\Delta n^{\neq} = -1$

$$p\Delta V^{\neq*} = -RT \tag{2.62b}$$

and

so that

$$E_A = \Delta H^{\neq*} + 2RT \tag{2.63b}$$

and

$$k = e^2\kappa \frac{kT}{h} \exp\left(-\frac{E_A}{RT}\right) \exp\left(\frac{\Delta S^{\neq*}}{R}\right). \tag{2.65}$$

2.6.2 Comparison and interpretations of the partition function form and the thermodynamic form of transition state theory with experimental data

The transition state theory expression given in terms of the experimental activation energy allows a direct comparison of the theoretical pre-exponential factor with the experimental pre-exponential factor, A.

The general form of the equation is

$$k = e^{1-\Delta n^{\neq}} \kappa \frac{kT}{h} \exp\left(-\frac{E_A}{RT}\right) \exp\left(\frac{\Delta S^{\neq*}}{R}\right) \tag{2.66}$$

where $1 - \Delta n^{\neq}$ is unity for unimolecular reactions, is two for bimolecular reactions and can be related to the molecularity of the reaction. The pre-exponential term is therefore

$$e^{1-\Delta n^{\neq}} \kappa \frac{kT}{h} \exp\left(\frac{\Delta S^{\neq*}}{R}\right).$$

This product corresponds directly to the A factor in the Arrhenius equation, and allows a theoretical calculation of the A factor once $\Delta S^{\neq*}$ has been found using partition functions derived from spectroscopic data and the potential energy surface for reaction.

$$A = e^{1-\Delta n^{\neq}} \kappa \frac{kT}{h} \exp\left(\frac{\Delta S^{\neq*}}{R}\right). \tag{2.67}$$

Table 2.7   Values of A and $\Delta S^{\neq\bullet}$ calculated from transition state theory showing dependence on molecular complexity

Reaction	$\log_{10} A_{calc}$ $(dm^3 mol^{-1} s^{-1})$	$\dfrac{\Delta S^{\neq\bullet}_{calc}}{JK^{-1} mol^{-1}}$
atom + diatomic molecule		
$H\cdot + H_2 \rightarrow H_2 + H\cdot$	10.7	-57
$Br\cdot + H_2 \rightarrow HBr + H\cdot$	11.1	-49
atomic + polyatomic molecule		
$H\cdot + CH_4 \rightarrow H_2 + CH_3$	10.3	-64
$H\cdot + C_2H_6 \rightarrow H_2 + C_2H_5\cdot$	10.1	-68
small molecule + small molecule		
$NO + Cl_2 \rightarrow ClNO + Cl\cdot$	9.1	-87
$2ClO\cdot \rightarrow Cl_2 + O_2$	7.5	-118
$NO_2 + CO \rightarrow NO + CO_2$	9.8	-74
$CH_3\cdot + H_2 \rightarrow CH_4 + H\cdot$	9.2	-85
small molecule + larger molecule		
$CH_3\cdot + CH_3COCH_3 \rightarrow CH_4 + \cdot CH_2COCH_3$	8.0	-108
larger molecule + larger molecule		
$C_4H_6 + C_2H_4 \rightarrow$ cyclo C_6H_{10}	7.8	-112

The A factor, therefore, explicitly reflects any trend in $\Delta S^{\neq\bullet}$ with reaction type. Table 2.7 tabulates values of $\Delta S^{\neq\bullet}$ for several bimolecular gas phase reactions. For these reactions $\Delta S^{\neq\bullet}$ is taken to reflect the complexity of the reacting molecules and activated complexes. Log A and $\Delta S^{\neq\bullet}$ parallel each other, and decrease as the complexity of the reacting molecules increases. This is particularly so with the Diels-Alder reactions where the discrepancy with collision theory is large, and the "p" factor is of the order of 10^{-5} to 10^{-6}.

This is an alternative, but equivalent, procedure to that discussed in Section 2.5 where the experimental A factor corresponded to the quotient

$$\kappa \frac{kT}{h} \frac{Q^{\neq\bullet}}{Q_X Q_{YZ}}$$

if E_o found from the potential energy surface is taken to correspond to the experimental activation energy, E_A.

The A factor found from Equation 2.67 involving $\Delta S^{\neq\bullet}$ corresponds to the data given in Table 2.3 where the very good agreement between experiment and theory is demonstrated.

An experimental $\Delta S^{\neq\bullet}$ can be found from the graph of ln k/T against $1/T$, and gives a direct comparison with the theoretical value of $\Delta S^{\neq\bullet}$. Again good agreement is found between theory and experiment, Table 2.8. But this would be expected to be the case.

If there is good agreement between theoretical and experimental A factors, then there must also be good agreement between the theoretical and experimental $\Delta S^{\neq*}$ values.

Table 2.8 Comparison of $\Delta S^{\neq*}$ values calculated from transition state theory with experimental $\Delta S^{\neq*}$ values

Reaction	$\dfrac{\Delta S^{\neq*}_{calc}}{JK^{-1}mol^{-1}}$	$\dfrac{\Delta S^{\neq*}_{obs}}{JK^{-1}mol^{-1}}$
$NO_2 + CO \rightarrow NO + CO_2$	-74	-93 to -68
$NO + O_3 \rightarrow NO_2 + O_2$	-97	-97 to -87
$2ClO\cdot \rightarrow Cl_2 + O_2$	-118	-120 to -110
$C_4H_6 + C_2H_4 \rightarrow cycloC_6H_{10}$	-112	-118
$2 cyclopentadiene \rightarrow dimer$	-120	-160 to -145

The "p" factor calculated from experiment can be compared with theory using $\Delta S^{\neq*}$ values. "p" gives the deviation of the observed A factor from the collision number, Z, and experiment shows that the magnitude of p decreases with increasing complexity of reaction type, Tables 2.7 and 2.8. As shown earlier, Section 2.5.1 the transition state theory analogue of the collision number is given by

$$\kappa \frac{kT}{h} \frac{f_r^2}{q_t^3}.$$

The deviation of the A factor from Z can be translated into transition state theory terms. The "p" factor is then given by the quotient

$$\kappa \frac{kT}{h} \frac{Q^{\neq*}}{Q_X Q_{YZ}} \bigg/ \kappa \frac{kT}{h} \frac{f_r^2}{q_t^3}.$$

It can be shown that

$$\frac{Q^{\neq*}}{Q_X Q_{YZ}} = e^{1-\Delta n^*} \exp\left(\frac{\Delta S^{\neq*}}{R}\right) \tag{2.68}$$

for reaction between molecules X and YZ. It can also be shown that

$$\frac{f_r^2}{q_t^3} = e^{1-\Delta n^*} \exp\left(\frac{\Delta S^{\neq*}_{A-A}}{R}\right) \tag{2.69}$$

for reaction between two atoms. It then follows that the "p" factor is given by the quotient

$$\exp\left(\frac{\Delta S^{\neq*}}{R}\right) \bigg/ \exp\left(\frac{\Delta S^{\neq*}_{A-A}}{R}\right) = \exp\left(\frac{\Delta S^{\neq*}}{R} - \frac{\Delta S^{\neq*}_{A-A}}{R}\right). \tag{2.70}$$

The "p" factor for any reaction can be expressed in terms of the difference between the calculated ΔS^{**} for the given reaction and ΔS^{**}_{A-A} calculated for the specific reaction of two atoms. Since the term, ΔS^{**}_{A-A}, is a constant whatever reaction is being studied, it also follows that trends in the observed "p" factor with complexity of reaction type can be compared directly with ΔS^{**} for the various reactions.

The comparisons between predicted behaviour and experimental observations

A factors reflect trends in $Q^{**}/Q_X Q_{YZ}$
A factors reflect trends in ΔS^{**}
p factors reflect trends in $Q^{**}/Q_X Q_{YZ}$
p factors reflect trends in ΔS^{**}

are all alternative, but equivalent, ways at looking at the implications of comparison of theory with experiment.

The inadequacy of collision theory lies in the neglect of the internal structure of reactants and activated complex, and in particular the lack of recognition of the fact that there is loss of rotational degrees of freedom in forming the activated complex. Transition state theory considers this explicitly.

The trend in A factors or p factors with complexity can be given as an entropy effect. For reaction between two complex molecules giving a non-linear activated complex there is a loss of three translational and three rotational degrees of freedom, and a gain of three vibrational degrees of freedom. The entropy associated with vibration is much less than that for rotation ($f_v < f_r$), consequently ΔS^{**} is expected to be more negative the more complex the molecule, Table 2.9.

Table 2.9   Typical approximate values of contributions entering into a ΔS^{**}

(i) Translational terms are based on a standard concentration taken to be 1 mol dm^{-3}
(ii) All values depend somewhat on temperature. Translational terms depend somewhat on temperature, and rotational terms on moments of inertia.

Translation:	ca 125 to 175 J mol^{-1} K^{-1}
	ca 40 to 60 J mol^{-1} K^{-1} per degree of freedom
Rotation:	ca 50 to 100 J mol^{-1} K^{-1} for linear molecules
	ca 75 to 150 J mol^{-1} K^{-1} for non-linear molecules
	ca 25 to 50 J mol^{-1} K^{-1} per degree of freedom
Vibration:	per normal mode
	from nearly zero, for high frequencies at normal temperatures, up to ca 15 J mol^{-1} K^{-1} for low frequencies at higher temperatures.

The examples given in Tables 2.3 to 2.9 are all for gas phase bimolecular reactions. Transition state theory is also applicable to the high pressure first order region of a unimolecular reaction, Table 2.10, and to the low pressure third order region of a termolecular reaction.

Table 2.10 Unimolecular decompositions

Reaction			$\log_{10}A$ A/s^{-1}	$\dfrac{E_A}{k\,Jmol^{-1}}$
cyclobutane	$\rightarrow$	$2C_2H_4$	15.6	261
C_2H_6	$\rightarrow$	$2CH_3\cdot$	17.4	384
$(CH_3)_3COOC(CH_3)_3$	$\rightarrow$	$2(CH_3)_3CO\cdot$	15.6	156
N_2O_5	$\rightarrow$	$NO_2 + NO_3\cdot$	14.8	88
cyclopropane	$\rightarrow$	propene	16.4	272
CH_3NC	$\rightarrow$	CH_3CN	13.6	160
$C_2H_5\cdot$	$\rightarrow$	$C_2H_4 + H\cdot$	13.0	167

2.7.1 Transition state theory applied to unimolecular reactions

The transition state theory expression for the first order region is

$$k = \kappa \frac{kT}{h} \frac{Q^{\neq *}}{Q_A} \exp\left(-\frac{E_0}{RT}\right).$$

(2.71)

The pre-exponential factor A is given by the terms

$$\kappa \frac{kT}{h} \frac{Q^{\neq *}}{Q_A}$$

if E_A and E_0 are taken to be equal.

The calculation of Q_A for the reactant A is a straightforward statistical mechanical calculation, but, as for bimolecular reactions, the structure of the activated complex must be found from the potential energy surface; or if this is not known sufficiently accurately, then the structure has to be guessed at or inferred from a comparison with molecules of comparable structure.

As before, the partition functions are split up into terms for translation, vibration and rotation. If the translational terms are taken to be similar for reactant and activated complex, these can be cancelled out leaving $Q^{\neq *}/Q_A$ as a ratio of partition functions for rotation and vibration.

2.7.2 Pre-exponential factors for reaction where the activated complex resembles molecule A

In a unimolecular reaction the activated complex has the same number of atoms as the reactant, and if the activated complex resembles the reactant molecule then the bond lengths in both are similar giving comparable moments of inertia. Both A^* and A will then have comparable rotational partition functions, and these will cancel out in the ratio $Q^{\neq *}/Q_A$, leaving the ratio to be made up of vibrational partition functions for activated complex and reactant.

In the formation of the activated complex one of the normal modes of vibration degenerates into a free translation in the activated complex, leaving the activated complex with one normal mode of vibration less than the reactant molecule. If the activated complex resembles the reactant then the frequencies for the normal modes of vibration in the activated complex will resemble those in the reactant, and the partition functions for these normal modes will be comparable, and will cancel each other out when the ratio Q^{**}/Q_A is formulated. This leaves the normal mode of vibration in the reactant which degenerates into the free translation in the activated complex still unaccounted for, and this partition function will remain uncancelled in the denominator of Q^{**}/Q_A. The corresponding free translation in the activated complex does not contribute to Q^{**} since it has already been incorporated at the earlier stage in the derivation of the working equation of transition state theory. This leaves

$$\text{predicted} \quad A = \kappa \frac{kT}{h} \frac{1}{f_v}.$$

$$(2.72)$$

Using typical values for f_v gives predicted values of A lying in the range 10^{14} s^{-1} to 10^{13} s^{-1}.

Many reactions have pre-exponential terms of these magnitudes, Table 2.10.

2.7.3   Pre-exponential factors for reaction where the activated complex differs considerably from reactant A

When the structure of the activated complex is very much altered from that of the reactant molecule A, the situation is now very different. For instance in a dissociation reaction such as:

$$C_2H_6 \quad \rightarrow \quad CH_3\cdot + CH_3\cdot$$

the C-C bond in the activated complex is likely to be considerably altered from what it was in the reactant, resulting in very different vibrational and rotational structures for the activated complex and reactant. There is likely to be an extension in the C-C bond so that

$$I^{\neq} \quad > \quad I_A$$

and hence $\quad f_r^{\neq} \quad > \quad f_{rA}$

giving

| the partition function for rotation in the activated complex | > | the partition function for rotation in the reactant |

Vibrations in the activated complex are "looser" than in reactant C_2H_6, and some may even become so distorted that they approximate to free rotations. These altered modes cause

$$v_{\neq} \quad < \quad v_A$$

so that

$$f_v^{\neq} \quad > \quad f_{vA}$$

giving

| the partition function for vibration in the activated complex | | the partition function for vibration in the reactant |

When typical values for f_r and f_v are taken, the values of A are found to lie in the range 10^{12} to 10^{18} s^{-1}.

These high values have also been found experimentally, see Table 2.10, and in the example chosen

$$C_2H_6 \quad \rightarrow \quad CH_3\cdot + CH_3\cdot$$

the value of A is 10^{17} s^{-1}.

High values of the pre-exponential factor in the high pressure region of a unimolecular decomposition can be used to infer a very "loose" structure for the activated complex, and low values imply a more "rigid" structure.

2.7.4 Transition state theory applied to termolecular reactions

The reactions of nitric oxide are thought to have a termolecular mechanism, for example

$$2NO + Cl_2 \qquad \rightarrow \qquad 2NOCl$$

Transition state theory gives

$$k = \kappa \frac{kT}{h} \frac{Q^{\neq*}}{Q_{NO}^2 Q_{Cl_2}} \exp\left(-\frac{E_0}{RT}\right).$$

(2.73)

An approximate calculation using the same values for each molecular partition function contributing to each Q allows cancellation of partition functions between activated complex and reactants. An order of magnitude value for the pre-exponential term can be found by substituting typical values for the molecular partition functions left in the ratio $Q^{\neq*}/Q^2_{NO} Q_{Cl_2}$ after cancellation. Doing this gives

$$\kappa \frac{kT}{h} \frac{Q^{\neq*}}{Q_{NO}^2 Q_{Cl_2}}$$

to be in the range 10^3 to 10^5 mol^{-2} dm^6 s^{-1}.

Agreement with experiment is good considering how crude the treatment is. The calculation is really only valid if partition functions for the reactants and activated complex are similar, and this is not likely. For instance, the molecular partition function for vibrations in the activated complex is likely to be different from those in the reactants.

A collision theory calculation of the pre-exponential factor is out by a factor of 10^{-5} when compared with the observed value, so even the simplified transition state theory expression is superior to collision theory for reactions similar to the NO/Cl$_2$ reaction.

One of the most interesting feature of suspected termolecular reactions is their zero or negative activation energies. Yet again transition state theory is superior, since there is no way in which collision theory can explain negative activation energies.

The dependence on temperature of each type of partition function is known. For the nitric oxide reactions, three linear molecules give a non-linear activated complex, and the dependence of the pre-exponential factor on temperature is $T^{-7/2}$.

The partition functions can be expressed in terms of a temperature independent, and a temperature dependent term giving

$$k = CT^n \exp\left(-\frac{E_0}{RT}\right).$$

(2.74)

In this expression C is the temperature independent part of the pre-exponential factor, and $n = -3.5$ for the reactions of nitric oxide.

This can be used to demonstrate that negative activation energies follow as a direct consequence of the temperature dependencies of the partition functions.

$$\ln k = \ln C + n \ln T - \frac{E_0}{RT}$$

(2.75)

$$\frac{d \ln k}{dT} = \frac{n}{T} + \frac{E_0}{RT^2}$$

$$= \frac{E_A}{RT^2}$$

(2.76)

$$E_A = nRT + E_0$$

(2.77)

where E_A is the experimental activation energy, and E_0 is the value found from the potential energy surface.

If E_0 is small and n is negative, then E_A can be negative. For instance, at 300 K, and for $n = -3.5$.

$$E_A = -8.7 + E_0$$

and so if E_0 is small E_A can be negative.

3

Advanced Transition State Theory

3.1 The assumptions of transition state theory reassessed

3.1.1 Assumption 1

The assumption is that once the unit (X..Y..Z) enters into the critical configuration it must continue on in the same direction (Figure 3.1).

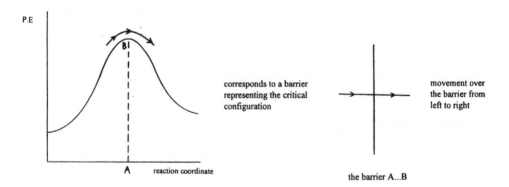

Figure 3.1 **A potential energy profile for reaction in which there is a single crossing of the potential energy barrier.**

However, if the unit did recross the barrier and moved back down into the entrance valley the situation would be given by Figure 3.2.

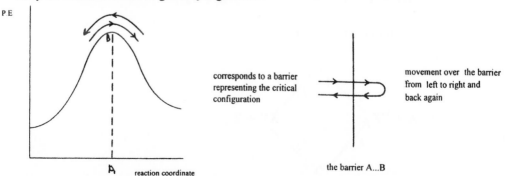

Figure 3.2 **A potential energy profile for reaction in which there is a recrossing back across the potential energy barrier.**

The net effect of recrossing the barrier in the opposite direction is zero passage through the critical configuration, but transition state theory would count this as one crossing from left to right, with the result that the calculated rate would be too high.

Similarly Figure 3.3 gives other situations.

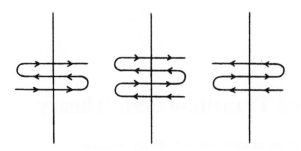

Figure 3.3 A diagram showing several recrossings of the potential energy barrier.

These diagrams show that if multiple crossings occur then transition state theory gives too high a rate. This can be taken care of by the transmission coefficient κ, or by variational transition state theory, see later.

3.1.2 Assumption 2

This relates to the assumption of equilibrium between reactants and activated complexes. Two situations can be discussed.

A. Single crossings through the critical configuration

Here reactants form activated complexes which can only move on to products. A forward moving complex is $\neq_f$, and one which started as products, and hence is a backward moving complex, is $\neq_b$.

$$X + YZ \xrightarrow{k_1} \neq_f \xrightarrow{k_2} XY + Z$$

$$XY + Z \xrightarrow{k_3} \neq_b \xrightarrow{k_4} X + YZ$$

These are part of a single scheme

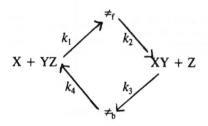

which itself is a more detailed description for the reversible reaction sequence.

$$X + YZ \underset{k_4}{\overset{k_1}{\rightleftharpoons}} \neq \underset{k_3}{\overset{k_2}{\rightleftharpoons}} XY + Z$$

in which the activated complexes $\neq$ are not distinguished as forward moving, $\neq_f$, and backward moving, $\neq_b$, complexes.

Situation 1: *Equilibrium is assumed for single crossings*
At equilibrium all concentrations must remain constant. In particular, the forward moving and backward moving complexes are at constant concentration, and define an equilibrium constant

$$\frac{[\neq_f]}{[\neq_b]} = K. \tag{3.1}$$

K must have a value of unity, a conclusion easiest seen by considering the statistical mechanical expression for the equilibrium constant

$$K = \frac{Q_{\neq f}}{Q_{\neq b}} \exp\left(-\frac{\Delta U_0}{RT}\right). \tag{3.2}$$

Since the only distinction between the activated complexes is that one is moving forward along the reaction coordinate and the other moving backward, the partition functions must be the same, and ΔU_0 is zero.

$$\therefore \quad \frac{[\neq_f]}{[\neq_b]} = 1$$

and since

$$[\neq_f] + [\neq_b] = [\neq] \tag{3.3}$$

then

$$[\neq_f] = [\neq_b] = \tfrac{1}{2}[\neq] \tag{3.4}$$

and at equilibrium

$$\frac{d[\neq_f]_{eq}}{dt} = 0. \tag{3.5}$$

Since

$$\frac{d[\neq_f]_{eq}}{dt} = k_1[X][YZ] - k_2[\neq_f]_{eq} = 0 \tag{3.6}$$

then

$$[\neq_f]_{eq} = \frac{k_1}{k_2}[X][YZ] \tag{3.7}$$

Likewise

$$[\neq_b]_{eq} = \frac{k_3}{k_4}[XY][Z]$$

(3.8)

and since

$$[\neq_f]_{eq} = [\neq_b]_{eq}$$

(3.9)

then

$$\frac{k_1}{k_2}[X][YZ] = \frac{k_3}{k_4}[XY][Z]$$

(3.10)

Situation 2: *A steady state is assumed for single crossings*
Now

$$\frac{d[\neq_f]_{ss}}{dt} = 0$$

(3.11)

giving

$$[\neq_f]_{ss} = \frac{k_1}{k_2}[X][YZ]$$

(3.12)

Likewise

$$\frac{d[\neq_b]_{ss}}{dt} = 0$$

(3.13)

giving

$$[\neq_b]_{ss} = \frac{k_3}{k_4}[XY][Z]$$

(3.14)

But, for single crossings at equilibrium

$$[\neq_f]_{eq} = \frac{k_1}{k_2}[X][YZ]$$

(3.15)

and

$$[\neq_b]_{eq} = \frac{k_3}{k_4}[XY][Z]$$

(3.16)

and so

$$[\neq_f]_{ss} = [\neq_f]_{eq}$$

(3.17)

$$[\neq_b]_{ss} = [\neq_b]_{eq}$$

(3.18)

from which it can be argued that the steady state situation can be treated as though it were an equilibrium situation, *provided the crossings are single*. And so the equilibrium assumption of transition state theory is valid, provided the crossings are single.

B. Multiple crossings through the critical configuration
The situation now is entirely altered. Forward moving complexes can turn into backward moving complexes, and vice-versa. And so the splitting up of the activated complexes into the two types of "forward moving only" and "backward moving only" is no longer possible, and a single type of activated complex must be considered.

$$X + YZ \underset{k_{-1}}{\overset{k_1}{\rightleftharpoons}} \neq \underset{k_{-2}}{\overset{k_2}{\rightleftharpoons}} XY + Z$$

where $\neq$ covers all types of activated complexes. In this scheme

$$\frac{d[\neq]}{dt} = k_1[X][YZ] - k_{-1}[\neq] - k_2[\neq] + k_{-2}[XY][Z] = 0 \qquad (3.19)$$

and so

$$[\neq] = \frac{k_1[X][YZ] + k_{-2}[XY][Z]}{k_{-1} + k_2} \qquad (3.20)$$

and these equations hold for both the equilibrium and steady state situations.

Situation 1: *Equilibrium is assumed for multiple crossings*
The equilibrium constant, K, for the *overall* reaction

$$X + YZ \rightleftharpoons XY + Z$$

is

$$K = \frac{[XY][Z]}{[X][YZ]} \qquad (3.21)$$

Since

$$[\neq]_{eq} = \frac{k_1[X][YZ] + k_{-2}[XY][Z]}{k_{-1} + k_2} \qquad (3.22)$$

$$\therefore \quad \frac{[\neq]_{eq}}{[X][YZ]} = \frac{k_1 \dfrac{[X][YZ]}{[X][YZ]} + k_{-2} \dfrac{[XY][Z]}{[X][YZ]}}{k_{-1} + k_2} \qquad (3.23)$$

$$= \frac{k_1 + k_{-2}K}{k_{-1} + k_2} \qquad (3.24)$$

$$\therefore \quad [\neq]_{eq} = \frac{k_1 + k_{-2}K}{k_{-1} + k_2}[X][YZ] \qquad (3.25)$$

The steady state expression for $[\neq]_{ss}$ is

$$[\neq]_{ss} = \frac{k_1[X][YZ]}{k_{-1} + k_2} + \frac{k_{-2}[XY][Z]}{k_{-1} + k_2} \qquad (3.26)$$

$$\qquad (3.27)$$

$$\therefore \quad [\neq]_{ss} \neq [\neq]_{eq}.$$

This argument shows that for *multiple* crossings the steady state concentration of activated complexes is no longer equal to the equilibrium concentration, and this is totally different from the result when *single* crossings only are allowed.

Situation 2: *A steady state is assumed for multiple crossings*
The differential equation is

$$\frac{d[\neq]_{ss}}{dt} = k_1[X][YZ] - k_{-1}[\neq] - k_2[\neq] + k_{-2}[XY][Z] = 0$$

(3.28)

As shown above equilibrium is *not* set up for multiple crossings and only a steady state situation is allowed. Hence the transition state theory given previously will not hold if multiple crossings are allowed.

The argument can be taken further, and the total forward rate calculated for various situations.

1. If k_{-1} and k_{-2} are zero, the forward rate is equal to the rate at equilibrium. This case corresponds to no recrossings allowed.
2. If k_{-1} and k_{-2} are very small, the forward rate is only slightly less than the value calculated on the assumption of equilibrium.
3. For all other values of k_{-1} and k_{-2}, the equilibrium assumption is invalid.

Therefore, in general the equilibrium assumption of transition state theory is invalid if there are multiple crossings.

3.1.3 Assumption 3
In transition state theory, motion through the critical configuration is singled out as special, and treated as though it were a free translation independent of the other internal movements of the activated complex. This is an approximation, for instance motion along the reaction coordinate can be affected by components normal to the reaction coordinate.

However, since this is a classical treatment the approximations are likely to be valid, but in exact treatments motion over the potential energy surface is quantised, and for such a situation it is not legitimate to consider motion over the barrier as independent of the other motions of the activated complex.

3.1.4 Assumption 4
Quantum effects manifest themselves in other ways. Reaction treated classically requires the unit (X..Y..Z) to pass over the top of the col. However, if movement on the surface is quantised, there is a finite probability that the unit (X..Y..Z) can move from one point on the reactant side of the barrier to another point on the product side without passing over the barrier. This is tunnelling, and reaction has occurred without the reactants having the critical energy, resulting in a transmission coefficient of greater than unity. Tunnelling is more likely when the barrier is thin and low, and for reactions of light mass species such as H, H^+, H^- or electrons.

At present there is no complete purely quantum theory, and the difficulties in producing one are considerable. In a quantum version, classical ideas will no longer be valid. For instance, motion over the potential energy surface must be quantised which requires that motion along the reaction coordinate cannot be separated from other internal motions. The partition function for the activated complex must be quantised, and so the partition function cannot be split into a free translation plus other partition functions. Quantisation of the partition function is vital for vibration, though a classical version for translation and rotation may sometimes be adequate, but only if the temperature and moments of inertia are both numerically high enough.

3.2 Symmetry Numbers and Statistical Factors

3.2.1 Symmetry Numbers
Rotational symmetry requires the partition function to be divided by a symmetry number, found by counting up the number of different, but equivalent, ways of arranging the molecule on rotation. For example, there are two ways of arranging each of Cl_2, H_2 and H_2O giving a symmetry number of two for each

```
    1      2              2      1
    Cl  —  Cl             Cl  —  Cl
    linear

    1      2              2      1
    H   —  H              H   —  H
    linear
```

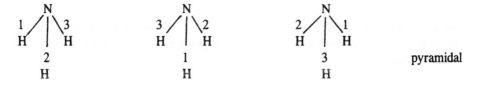

```
        O.                        O.
    H /     \ H            H /        \ H          bent
    1         2            2           1
```

The shape of the molecule is also important. For instance NH_3 is non planar with a symmetry number of three.

```
        N                     N                     N
    1 / | \ 3             3 / | \ 2             2 / | \ 1
    H   |   H             H   |   H             H   |   H
        2                     1                     3         pyramidal
        H                     H                     H
```

Turning over non-planar molecules such as NH_3 does not generate any more distinct arrangements.

$CH_3\cdot$ is planar with a symmetry number of six. Turning over the following three arrangements

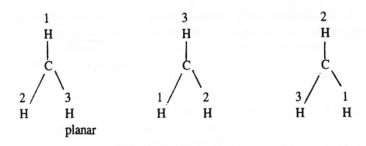

planar

generates three more distinct arrangements

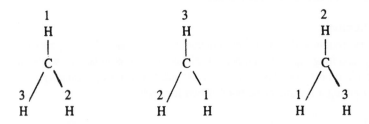

These symmetry numbers must be taken into account when formulating statistical mechanical equilibrium constants.

The equilibrium constant for the reaction

$$A + B \;\rightleftharpoons\; C + D$$

can be written as

$$K = \frac{Q_C Q_D}{Q_A Q_B} \exp\left(-\frac{\Delta U_0}{RT}\right) \tag{3.29}$$

or as

$$K = \frac{\sigma_A \sigma_B}{\sigma_C \sigma_D} \frac{Q_C^0 Q_D^0}{Q_A^0 Q_B^0} \exp\left(-\frac{\Delta U_0}{RT}\right) \tag{3.30}$$

where the symmetry numbers have been removed from the standard partition function, Q, to give a partition function Q^0 which has no symmetry number, σ, in the rotational partition function.

3.2.2 Statistical factors
An alternative procedure is to use statistical factors, l for the forward reaction and r for the back reaction.

$$K = \frac{l}{r} \frac{Q_C^0 Q_D^0}{Q_A^0 Q_B^0} \exp\left(-\frac{\Delta U_0}{RT}\right). \tag{3.31}$$

Expressions using l and r are simpler to handle and less likely to lead to error than those using symmetry numbers. However, these statistical factors have to be defined separately for reactions between unlike molecules and those between like molecules.

Unlike molecules   A + B

The statistical factor is equal to the number of ***distinct sets*** of products which would result if all like atoms were distinguishable and were labelled.

For instance, reaction between H_2 and $Cl\cdot$, between H_2 and $H\cdot$ and between N_2 and O_2 leads to two sets of products in each case

(a)

$$\overset{1}{H} - Cl + \overset{2}{H}\cdot$$

$$\overset{1}{H} - \overset{2}{H} + Cl\cdot$$

$$\overset{2}{H} - Cl + \overset{1}{H}\cdot$$

two distinct sets of products $\therefore l = 2$

(b)

$$\overset{1}{H} - \overset{3}{H} + \overset{2}{H}\cdot$$

$$\overset{1}{H} - \overset{2}{H} + \overset{3}{H}\cdot$$

$$\overset{2}{H} - \overset{3}{H} + \overset{1}{H}\cdot$$

two distinct sets of products $\therefore l = 2$

(c)

$$\overset{1}{N} = \overset{3}{O} + \overset{2}{N} = \overset{4}{O}$$

$$\overset{1}{N} \equiv \overset{2}{N} + \overset{3}{O} = \overset{4}{O}$$

$$\overset{1}{N} = \overset{4}{O} + \overset{2}{N} = \overset{3}{O}$$

two distinct sets of products $\therefore l = 2$

Like molecules   A + A

The statistical factor is now given by the number of ***distinct sets*** of products which would result if all like atoms were distinguishable and labelled, ***divided by two***.

For instance, reaction between two N = O molecules gives one set of products only

$$
\begin{array}{cccc}
1 & 2 & 3 & 4 \\
N & = O & + N & = O
\end{array}
\quad \rightarrow \quad
\begin{array}{cccc}
1 & 3 & 2 & 4 \\
N & \equiv N & + O & = O
\end{array}
$$

but the statistical factor is now the number of sets of products divided by two

$$\therefore l = \tfrac{1}{2}$$

3.2.3 Equilibrium constants using statistical factors

The statistical factors, l and r, for a reaction coming to equilibrium must first be worked out. For example, in the reaction

$$H_2 + I_2 \longrightarrow 2HI$$

the statistical factor for the forward reaction is two

$$
\begin{array}{cccc}
1 & 2 & 3 & 4 \\
H & - H & + I & - I
\end{array}
\Big\langle
\begin{array}{l}
\begin{array}{cccc}
1 & 3 & 2 & 4 \\
H & - I & + H & - I
\end{array} \\[1em]
\begin{array}{cccc}
1 & 4 & 2 & 3 \\
H & - I & + H & - I
\end{array}
\end{array}
$$

There are two distinct sets of products, and reaction is between unlike reactants

$$\therefore l = 2$$

However, the back reaction is between like molecules, and gives one set of products

$$
\begin{array}{cccc}
1 & 3 & 2 & 4 \\
H & - I & + H & - I
\end{array}
\quad \rightarrow \quad
\begin{array}{cccc}
1 & 2 & 3 & 4 \\
H & - H & + I & - I
\end{array}
$$

$$\therefore r = \tfrac{1}{2}$$

The equilibrium constant expression

$$K = \frac{l}{r} \frac{\left(Q^0_{HI}\right)^2}{Q^0_{H_2} Q^0_{I_2}} \exp\left(-\frac{\Delta U_0}{RT}\right)$$

$$= 4 \frac{\left(Q^0_{HI}\right)^2}{Q^0_{H_2} Q^0_{I_2}} \exp\left(-\frac{\Delta U_0}{RT}\right). \tag{3.32}$$

The decomposition of O_3

$$O_3 \quad \rightarrow \quad O_2 + O\cdot$$

illustrates one other important point.

Both the forward and back reaction give three sets of products so that both l and r are equal to three.

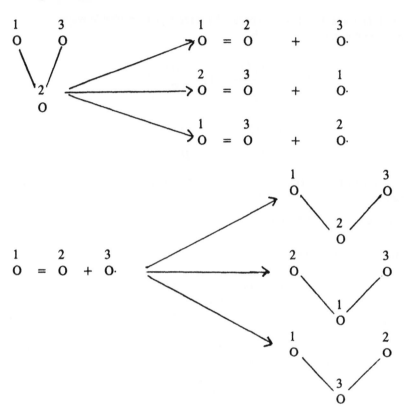

In the forward direction the products $\overset{1}{O} = \overset{3}{O} + \overset{2}{O}\cdot$ are included, while in the back reaction the product with the middle oxygen atom labelled 3 is included because a likely transition state is

Breakdown of this transition state is as likely to proceed by splitting of

$$\overset{1}{O} - \overset{2}{O} \text{ bond and the } \overset{2}{O} - \overset{3}{O} \text{ bonds}$$

as by splitting of

$$\begin{matrix} 1 & 2 & & 1 & 3 \\ \text{the O} & \text{---} & \text{O bond and the} & \text{O} & \text{---} & \text{O bonds} \end{matrix}$$

or by splitting of

$$\begin{matrix} 1 & 3 & & 2 & 3 \\ \text{the O} & \text{---} & \text{O bond and the} & \text{O} & \text{---} & \text{O bonds} \end{matrix}$$

Formation of the transition state also occurs by three equally probable ways. The equilibrium constant expression is

$$K = \frac{3 \, Q_{O_2}^0 Q_O^0}{3 \, Q_{O_3}^0} \exp\left(-\frac{\Delta U_0}{RT} \right)$$

$$= \frac{Q_{O_2}^0 Q_O^0}{Q_{O_3}^0} \exp\left(-\frac{\Delta U_0}{RT} \right). \tag{3.33}$$

Reactions involving molecules with very high symmetry can have very large statistical factors.

$$PCl_5 \rightleftharpoons PCl_3 + Cl_2$$

Here $l = 20$ and $r = 20$ so that

$$K = \frac{Q_{PCl_3}^0 Q_{Cl_2}^0}{Q_{PCl_5}^0} \exp\left(-\frac{\Delta U_0}{RT} \right). \tag{3.34}$$

Special care must be taken when formulating statistical factors for reactions producing enantiomers. If reaction produces two enantiomers as in

$$H\cdot + Cl\,CF\,Br\cdot \rightleftharpoons CH\,Cl\,F\,Br$$

then the statistical mechanical expression using symmetry numbers gives an equilibrium constant which corresponds to formation of one enantiomer only. The expression using statistical factors, l and r, gives an equilibrium constant for formation of the two enantiomers where $l = 2$ and $r = 1$. Hence the two equilibrium constant formulations will differ by a factor of two.

3.2.4 Rate constants in terms of statistical factors

Rates are also handled much more safely using statistical factors rather than symmetry numbers. Reaction is treated similarly to equilibrium, only now equilibrium is between reactants and activated complex, and between products and activated complex.

For the reaction

$$A + B \quad \rightarrow \quad C + D$$

the activated complex is $AB^{\ddagger}$ and the two equilibria which must be considered are

$$A + B \rightleftharpoons AB^{\ddagger}$$
$$C + D \rightleftharpoons AB^{\ddagger}$$

Each equilibrium has statistical factors for the forward and back reactions, and the directions of these steps are always defined in terms of *formation of $AB^{\ddagger}$ being the forward step*.

In the equilibrium

$$A + B \underset{r^{\ddagger}_1}{\overset{l^{\ddagger}_1}{\rightleftharpoons}} AB^{\ddagger}$$

$l^{\ddagger}_1$ represents the statistical factor for the forward direction of forming the activated complex $AB^{\ddagger}$, and $r^{\ddagger}_1$ is the statistical factor for removal of $AB^{\ddagger}$ reforming A and B.

In the equilibrium

$$C + D \underset{r^{\ddagger}_{-1}}{\overset{l^{\ddagger}_{-1}}{\rightleftharpoons}} AB^{\ddagger}$$

$l^{\ddagger}_{-1}$ represents the statistical factor for the forward direction of forming $AB^{\ddagger}$ while $r^{\ddagger}_{-1}$ is the statistical factor for removal of $AB^{\ddagger}$ forming C and D.

In the overall reaction of converting A and B into C and D, the statistical factors appear as follows

$$A + B \underset{r^{\ddagger}_1}{\overset{l^{\ddagger}_1}{\rightleftharpoons}} AB^{\ddagger} \underset{l^{\ddagger}_{-1}}{\overset{r^{\ddagger}_{-1}}{\rightleftharpoons}} C + D$$

In transition state theory only one set of products and one set of reactants can be formed from the activated complex, otherwise there would be more than one entrance valley and more than one exit valley. This means that the statistical factors for removing $AB^{\ddagger}$ to products, and for removing $AB^{\ddagger}$ to reactants must both be unity so that

$$r^{\ddagger}_{-1} = 1 \qquad \text{and} \qquad r^{\ddagger}_1 = 1$$

In structural terms this restriction implies that the activated complex must be of sufficiently low symmetry that there are only two routes away from the activated complex. This results in a scheme

$$A + B \underset{r^{\ddagger}_1 = 1}{\overset{l^{\ddagger}_1}{\rightleftharpoons}} AB^{\ddagger} \underset{l^{\ddagger}_{-1}}{\overset{r^{\ddagger}_{-1} = 1}{\rightleftharpoons}} C + D$$

When this is the case the equilibrium constant for formation of $AB^{\ddagger}$ from A and B is

$$K^{\neq*} = \frac{l_1^{\neq}}{r_1^{\neq} = 1} \frac{Q^{0\neq*}}{Q_A^0 Q_B^0} \exp\left(-\frac{E_0}{RT}\right)$$ (3.35)

giving

$$k = \kappa \frac{kT}{h} l_1^{\neq} \frac{Q^{0\neq*}}{Q_A^0 Q_B^0} \exp\left(-\frac{E_0}{RT}\right).$$ (3.36)

Since the rate expression deals with the forward steps only, the equilibrium involving formation of AB^* from C and D does not enter into the procedure.

The statistical factor $l_1^{\neq}$ must be found, and this is done by counting the number of activated complexes formed if similar atoms in the reactants and activated complexes are labelled. This will be explained by describing several cases.

1. In the reaction

$$H_2 + Cl\cdot \qquad \rightarrow \qquad HCl + H\cdot$$

formation of the activated complex from the reactants gives two distinct activated complexes

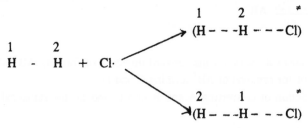

which results in a statistical factor $l_1^{\neq} = 2$ so that

$$k = \kappa \frac{kT}{h} 2 \frac{Q^{0\neq*}}{Q_{H_2}^0 Q_{Cl}^0} \exp\left(-\frac{E_0}{RT}\right).$$ (3.37)

2.   In the reaction

$$HO\cdot + H_2 \qquad \longrightarrow \qquad H_2O + H\cdot$$

two activated complexes can again be formed from the reactants.

$$
\begin{array}{ccc}
1 & 2 & 3 \\
H\text{-----}O\ + & H\text{----}H
\end{array}
$$

$$
\begin{array}{cccc}
1 & & 2 & 3 \; * \\
(H\text{- - -}O\text{- - -}H\text{- } \bullet \text{ -}H)
\end{array}
$$

$$
\begin{array}{cccc}
1 & & 3 & 2 \; * \\
(H\text{- - -}O\text{- - -}H\text{- - -}H)
\end{array}
$$

giving a statistical factor $l^{\neq}_1 = 2$, and rate constant

$$k = \kappa \frac{kT}{h} 2 \frac{Q^{0 \neq *}}{Q^0_{H_2} Q^0_{H\cdot}} \exp\left(-\frac{E_0}{RT}\right).$$

(3.38)

3.   However, in the H. abstraction reaction

$$H\cdot + CH_4 \longrightarrow H_2 + CH_3\cdot$$

four activated complexes can be found

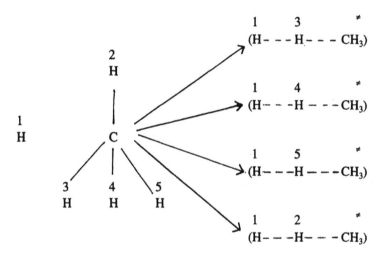

giving a statistical factor $l^{\neq}_1 = 4$, and rate constant

$$k = \kappa \frac{kT}{h} 4 \frac{Q^{0 \neq *}}{Q^0_{H\cdot} Q^0_{CH_4}} \exp\left(-\frac{E_0}{RT}\right).$$

(3.39)

4.   If the activated complex is chiral then the forward rate must include a statistical factor which reflects this.

3.2.5   Use of statistical factors in the prediction of possible structures for the activated complex

Statistical factors are very useful in excluding certain structures for an activated complex. Since the potential energy surface must have only one entrance and one exit valley, the structure of the activated complex must be such that this restriction can be satisfied. This, as mentioned earlier, implies that $r^{\neq}_{-1}$ and $r^{\neq}_{1}$ must both be unity and this in turn, means that the activated complex must have low symmetry.

1.   In the reaction

$$H\cdot + DOH \longrightarrow D\cdot + H_2O$$

a possible structure for the activated complex might be a pyramidal DOH_2

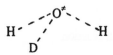

For this structure to be acceptable only one set of products can be formed from the activated complex, so that $r^{\neq}_{-1}$ equals unity. The pyramidal activated complex satisfies this condition.

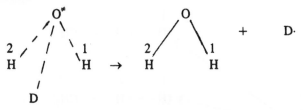

Return of the activated complex must also result in one set of reactants $H\cdot + DOH$ so that $r^{\neq}_{1}$ equals unity. However, the pyramidal structure would lead to two sets of reactants giving $r^{\neq}_{1} = 2$, and so a pyramidal $DOH_2^{\neq}$ is not an acceptable structure for the activated complex in this reaction

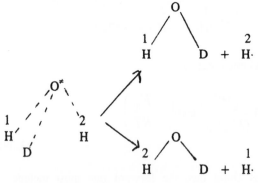

A planar activated complex could also be considered. Here one set of products would be formed satisfying the criterion that $r^{\neq}_{-1} = 1$

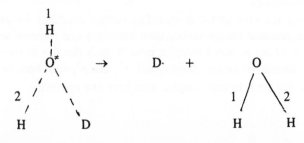

However, the planar activated complex is ruled out because $r^{\ddagger}_1 = 2$.

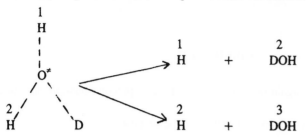

The highly unsymmetrical planar activated complex has one long OH bond and one short OH bond.

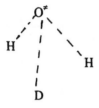

These can be labelled differently so that two activated complexes can be formed from the reactants giving $l^{\ddagger}_1 = 2$.

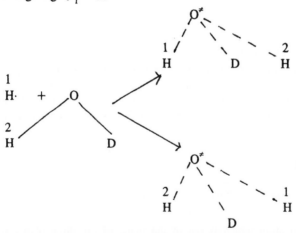

However, it is $r^{\ddagger}_{-1}$ and $r^{\ddagger}_1$ which are decisive in determining whether these unsymmetrical planar activated complexes are acceptable structures.

Since formation of product involves breaking of the O-D bond, only one set of products can be formed from either of the activated complexes. This will give $r^{\ddagger}_{-1} = 1$ for either activated complex.

Return of the activated complex to reactants involves breaking an O-H bond, and here the long O-H bond will break preferentially. This means that only one set of reactants can be formed from the activated complex, giving $r^{\ddagger}_1 = 1$. Hence an unsymmetrical planar activated complex satisfies both conditions $r^{\ddagger}_{-1} = 1$ and $r^{\ddagger}_1 = 1$. and so will be an acceptable structure in contrast to the planar or pyramidal structures where each OH bond has an equal chance of breaking.

2.   The isomerisation of cyclopropane is a well-documented reaction,

$$
\begin{array}{c}
CH_2 \\
\diagup \quad \diagdown \\
CH_2 \!\!-\!\!-\!\! CH_2
\end{array}
\quad \rightarrow \quad CH_3CH = CH_2
$$

and statistical factors have again been of use in helping to pinpoint possible structures for the activated complex.

For example, a planar symmetrical structure can be proposed.

This structure results in the following statistical factors

$$
\begin{array}{ll}
l^{*}_{1} = 3 & r^{*}_{1} = 1 \\
l^{*}_{-1} = 3 & r^{*}_{-1} = 2
\end{array}
$$

and is ruled out on the grounds that $r^{*}_{-1} = 2$ even although $r^{*}_{1} = 1$.

The structure which has to be proposed is one which can produce propene in one way only. The non-planar activated complex where the three carbon atoms lie in a planar triangle with one hydrogen out of the plane has very low symmetry, and gives both r^{*}_{-1} and r^{*}_{1} equal to unity. It is, therefore, a possible acceptable structure.

H^{*} is the hydrogen atom which is out of the plane of the three planar triangular carbons, and is the hydrogen atom which migrates from one of the carbons forming the double bond to the carbon atom which ends up with three hydrogen atoms.

3.3 Advances in transition state theory
At present this is a very active field with developments being pursued in several directions. Generalised transition state theory is one major area, and this looks at points along the reaction coordinate, other than at the col. In conventional transition state theory the transition state is at the col or saddle point on the potential energy surface, and the rate of reaction is calculated as the rate of passage through this col. At low temperatures or energies, reactive trajectories pass through the col and rarely return, but at high temperatures or energies the col may no longer determine the rate,

and generalised transition state theory becomes important. Other points along the
reaction coordinate are now looked at in detail, and generalised transition state theory
focuses on these regions. The terminology is now no longer "passage over the col",
instead, "crossing a dividing surface" is used.

The dividing surface straddles the reaction coordinate at right angles, and is defined
by a curve joining two equal potential energies, one on each side of the reaction
coordinate. The dividing surface cuts across the valley, and so the curve is concave
upward giving a potential well across the reaction coordinate, Figure 3.4.

The simplest representation of a dividing surface is a plane through the surface at
right angles to the reaction path, though sometimes it is simply drawn as a line,
Figures 3.5(i), 3.5(ii).

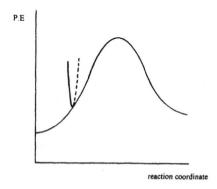

Figure 3.4 A dividing surface.

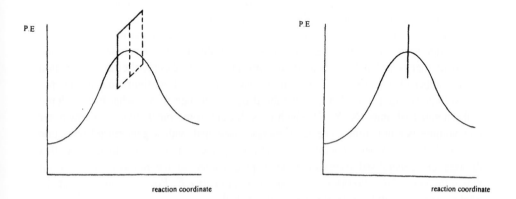

Figure 3.5(i), (ii)   A dividing surface at the col.

In Figures 3.5(i), 3.5(ii), the dividing surface is placed at the col, but it can be placed at
any position along the reaction coordinate, Figure 3.6(i), 3.6(ii).

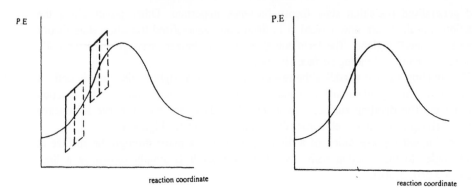

Figure 3.6(i), (ii) Dividing surfaces along the reaction coordinate.

Regions of the reaction coordinate at which a dividing surface is placed are called "generalised transition states", and these are always displaced from the col, and from now on the term transition state is reserved for a dividing surface at the col. Some of these generalised transition states will be of no significance to rate theory, but others will be of considerable importance, and may even be found to control the rate.

As defined earlier, the dividing surface is a plane perpendicular to the reaction coordinate, and the generalised transition state lies beneath the dividing surface. For the generalised transition states, the potential energy increases in both directions at right angles to the reaction coordinate, and so it is in a potential well in this direction. Along the reaction coordinate , displacement of the generalised transition state to lower positions on the reaction coordinate results in a decrease in potential energy, while displacement to higher positions results in an increase in potential energy. In consequence, the generalised transition state is only a particular configuration, but it is still radically different from the transition state at the critical configuration which sits at a potential energy maximum along the reaction coordinate.

Conventional transition state theory deals exclusively with the dividing surface at the col, while modern extensions deal mainly with generalised transition states. Further differences in terminology now appear. In transition state theory the minimum energy for reaction to occur is called the critical energy, in the modern developments it is the threshold energy which is often used. In conventional rate theory, the rate of reaction is the number of units (X..Y..Z) passing through the critical configuration, or col, per unit volume per unit time, but in generalised transition state theory what is considered is the number of units (X..Y..Z) which pass through some other dividing surface per unit volume per unit time. The total energy associated with a generalised transition state includes the kinetic energy of motion along the reaction coordinate. The flux through the generalised transition state is proportional to the number of states for which the *rest of the energy*[1] is less than the total energy E. For reaction to occur the total energy E has to be at least equal to the critical.

The theory calculates these quantities at various positions along the reaction coordinate where the dividing surfaces have been placed.

The generalised transition state which has the lowest number of states with the *rest of the energy* less than E, is associated with a flux minimum, and this corresponds to a

---

[1] i.e. the total energy minus the kinetic energy of motion along the reaction coordinate.

barrier or bottleneck to reaction. This arises because the rate of passage through the generalised transition state is proportional to this number of states, hence low rates correspond to barriers or bottlenecks which themselves correspond to flux minima. The lowest flux thus corresponds to the lowest rate of passage through a generalised transition state. Generalised transition states which have very large numbers of states are associated with flux maxima which can be thought of as wells. If the flux is calculated at every point along the reaction coordinate a sequence of barriers and wells may be found, in contrast to the smooth passage of conventional transition state theory.

The bottleneck or barrier corresponding to the lowest flux thus controls the rate of reaction, and this barrier can be anywhere along the reaction coordinate between the reactants and the activated complex. However, to achieve reaction the "reaction entity" *must also reach* and *pass through* the critical configuration.

The essential difference between traditional transition theory and generalised transition state theory can be summarised in the two statements:

a) In traditional transition state theory the critical configuration must be reached for reaction to occur, and the rate of passage through this critical configuration lying at the col defines the rate of reaction.

b) In generalised transition state theory the critical configuration at the col must be reached and passed through for reaction to occur, but now the rate is controlled by the rate of passage through the configuration associated with the flux minimum.

This type of calculation can be informative. If there is only one barrier to reaction and no wells, then reaction occurs by a "direct mechanism", but if there is a flux maximum bounded by two flux minima this corresponds to a well bounded by two barriers, and reaction then occurs by a "complex reaction". Predictions about the existence of wells can now be tested directly by molecular beam experiments, Chapters 10 and 13.

The simplest "complex" mechanism corresponds to two barriers with a well in between. The trajectory of an (X..Y..Z) unit can pass over the first barrier, enter the well, and possibly cross and recross the well many times before leaving it via either barrier. This will give an upper limit to the rate. The lower limit corresponds to a single or double crossing of the well, and this is equivalent to a direct mechanism. Here the entrance and exit barriers have flux minima almost equal to that for the flux maximum of the well. Entry from one barrier into the well is followed by exit through the other. Since the three fluxes are virtually coincident the difference from "direct" reaction is minimal.

One study carried out at various energies gave interesting results. For a particular potential energy surface and for energies near the threshold, the classical number of states showed a minimum near the col, indicating direct reaction with a barrier near the saddle point. As the energy increased, progressive changes were observed, until, at high energies, the flux minimum had changed to a flux maximum bounded by two flux minima on each side, indicating that there is a well with two barriers on each side. These properties correspond to a collision complex. This prediction has been picked up in molecular beam studies of some reactions.

Generalised transition states are also used in variational microcanonical and canonical transition state theory.

3.31   Variational transition state theory

Dividing surfaces are placed at various positions along the reaction coordinate, and the rate of passage through each dividing surface, or generalised transition state, is calculated. The dividing surface which gives the lowest rate is the nearest to the truth, provided tunnelling is negligible, and this rate is either equal to or greater than the true rate of reaction. In contrast, conventional transition state theory will only give an upper limit to the rate. Since the position of the dividing surface is varied, and the rate minimised, this technique is called variational transition state theory.

Multiple crossings through the transition state at the col have been shown to result in too high a rate, and the dividing surface gives a means of minimising this effect, Figure 3.7.

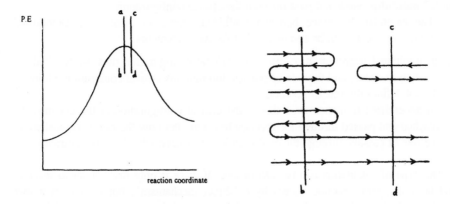

Figure 3.7   Use of movement of dividing surfaces to minimise the effect of multiple recrossings.

The dividing surface a-b cuts the reaction coordinate at the col, and the trajectory has many crossings. These can be minimised by moving the surface to c-d. Because there are now considerably fewer crossings than at a-b, the rate calculated at c-d will be less than that at a-b. Variational transition state theory actually minimises the rate itself, and, in doing so, automatically chooses a dividing surface with a minimum in the number of multiple crossings. The minimum rate gives the best estimate of the actual rate.

Variational transition state theory is a tool to enable the optimum dividing surface to be picked. Microcanonical and canonical transition state theory then utilise the result of the variational procedure to calculate the rate of reaction.

3.3.2   Microcanonical transition state theory

This is a theory which calculates the rate of reaction and the rate constant at one single energy. The reaction rate is found at energy E where E is greater than the threshold energy. But in the process of getting to this rate all energies up to E have to be considered. Energy will be distributed among all the various (X..Y..Z) units by a Maxwell-Boltzmann distribution if the equilibrium assumption holds, giving a spread in the values of the total energies of the various (X..Y..Z) units from zero upwards. These energies are given by horizontal planes through potential energy surfaces. The extent to which an (X..Y..Z) unit can move along the reaction coordinate is determined by its total energy, low energy units cannot get very far along the reaction coordinate while the ones with higher energy, though still below the threshold, can get much nearer the col. Only units (X..Y..Z) with total energy at least equal to the threshold can ever reach the col.

Microcanonical transition state theory is interested in what happens to all the dividing surfaces displaced from the col. All units (X..Y..Z) which are in a given reference generalised transition state have a total energy at least as great as the minimum energy which they must have to enter the generalised transition state. This total energy is given by a horizontal plane through the reference generalised transition state. However, units with energies above this minimum can still be resident in the generalised transition state. Energies below the horizontal plane are not possible for units (X..Y..Z) actually in the reference generalised transition state.

Attainment of the configuration of the reference generalised transition state automatically means possession of at least the minimum energy corresponding to the horizontal plane through the generalised transition state. Units (X..Y..Z) which have less total energy will lie in lower generalised transition states, but units (X..Y..Z) which have a greater total energy than that of the reference generalised transition state will either lie in the reference state, or in higher ones. So, a particular generalised transition state will contain units (X..Y..Z) which have energy equal to the minimum total energy associated with entry to the generalised transition state and which cannot move any further along the reaction coordinate, or units (X..Y..Z) with higher total energy which are just passing through, en route to higher generalised transition states. Units (X..Y..Z) with a given total energy will eventually pass through all the dividing surfaces up to the generalised transition state with a minimum energy entrance requirement equal to that of the unit (X..Y..Z). When a count is made of the units (X..Y..Z) in a particular generalised transition state, this will be made up of:

1. All units (X..Y..Z) which are actually in the generalised transition state and have just the minimum energy to get into it. There will be other units (X..Y..Z) with the appropriate energy which have not got to the appropriate configuration, but are still passing through the lower generalised transition states.
2. There are units (X.Y..Z) which are temporarily resident in the particular generalised transition state, but have enough energy to take them eventually to higher generalised transition states.

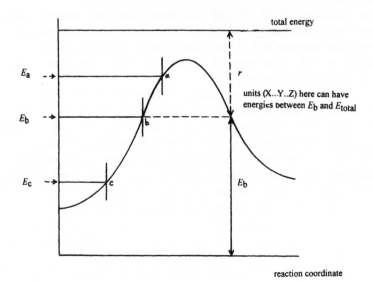

E_a, E_b, E_c are energies which represent the minimum
energy which a unit (X...Y...Z) must have before it can
enter the generalised transition states a, b, c respectively.

At b have units (X...Y...Z) with energy E_b and units
(X...Y...Z) with energy $>E_b$ which are en route to
positions further along the reaction coordinate. Units
(X...Y...Z) with energy $<E_b$ cannot get to the generalised
transition state b.

Figure 3.8 The use of dividing surfaces in generalised transition state theory.

Dividing surfaces are placed along the reaction coordinate, Figure 3.8, and the
classical number of states in each generalised transition state is calculated for each
energy equal to, or greater than, the minimum energy for entry into the particular
generalised transition state, and up to the value E of the total energy for which the
calculations are being made, that is, all units (X..Y..Z) lying in the range r in the
diagram above, Figure 3.8. This is so that the calculations include units (X..Y..Z)
with higher energies, but which are only passing through en route to higher generalised
transition states. This calculation is repeated at every dividing surface, and the dividing
surface which gives a minimum value to the number of states is identified. This then
pinpoints the region of the reaction coordinate which corresponds to a bottleneck or
barrier to reaction, so that passage through this generalised transition state will control
the rate of reaction. This is the dividing surface which will also give a minimum to the
rate and the rate constant, and this minimum rate is taken to be the best estimate to the
true rate of reaction.

Not all units (X..Y..Z) which contribute to this rate will actually attain reaction. Of
the units (X..Y..Z) which pass through the bottleneck at the critical dividing surface,
only those which have an energy which is greater than the threshold will actually be
able to go on to reaction. And so the theory will, in fact, over-estimate the rate of
reaction.

The rate and rate constant for the critical dividing surface has been found for reaction at one total energy E. This cannot give an overall rate for which all energies would have to be considered. But integration over all energies, assuming a Maxwell-Boltzmann distribution will give the overall or thermal rate constant.

$$k(T) = \int_0^\infty \frac{\Phi(E)}{Q(T)} \exp\left(-\frac{E}{RT}\right) k(E) \mathrm{d}E$$

(3.40)

where $\Phi(E)$ is the number of states per unit range of energy per unit volume for reactants. $k(E)$ is the microcanonical rate constant for one energy (E). $Q(T)$ is the partition function(s) for reactants.

Micro-canonical transition state theory is equivalent to vibrationally adiabatic theory where the initial quantum states of the unit $(X..Y..Z)$ are chosen, and the system remains in these states throughout, but where motion along the reaction coordinate is treated classically , see Quack and Troe, Chapter 5 pp128-143.

3.3 Canonical transition state theory
The thermal rate constant $k(T)$

$$k(T) = \int_0^\infty \frac{\Phi(E)}{Q(T)} \exp\left(-\frac{E}{RT}\right) k(E) \mathrm{d}E$$

(3.41)

is a rate constant averaged over all energies using a Maxwell-Boltzmann distribution, and is therefore, by definition, a canonical rate constant, in contrast to $k(E)$ which is a rate constant for one energy only, and so is, by definition, a microcanonical rate constant.

In canonical transition state theory $k(E)$ is worked out for one dividing surface just as in the microcanonical theory. From this, $k(T)$ is found by integration, as above, for that particular dividing surface. $k(E)$ is then worked out for a further series of dividing surfaces and the corresponding $k(T)$ values found. The minimum $k(T)$ is then found, and this is taken as the best estimate to the true rate.

It is not immediately clear from a verbal account just where and how the two minimising procedures differ. The following sets down this difference explicitly.

Both procedures start by providing values of $k(E)$ for a variety of dividing surfaces.

(a) Microcanonical theory:
This compares $k(E)$ for various dividing surfaces, and then minimises by taking the minimum value $k'(E)$ as the best $k(E)$. From this

$$k(T) = \int_0^\infty \frac{\Phi(E)}{Q(T)} \exp\left(-\frac{E}{RT}\right) k'(E) \mathrm{d}E$$

(3.42)

when $k'(E)$ may relate to different dividing surfaces for different values of E.

(b) Canonical theory:
This works out various $k(T)$ for a variety of dividing surfaces, and the minimum value
of $k'(T)$ is taken

$$k'(T) = \int_0^\infty \frac{\Phi(E)}{Q(T)} \exp\left(-\frac{E}{RT}\right) k(E) dE$$

(3.43)

where this involves a $k(E)$ which is not necessarily $k'(E)$. All $k(E)$ in this integral relate
to the same dividing surface.

Both the micro and canonical theories include units (X..Y..Z) which do not have the
threshold energy, and so can never move through the col and thence to products.
These contribute to $k(E)$, but since they do not result in reaction cause $k(T)$ to be
overestimated. They must therefore be omitted from the calculation. They are best
eliminated by integrating over all energies from E_0 (the threshold) to ∞ rather than
from 0 to ∞, that is including only units which can react.

$$k(T)\text{improved} = \int_{E_0}^\infty \frac{\Phi(E)}{Q(T)} \exp\left(-\frac{E}{RT}\right) k(E) dE.$$

(3.44)

This is called *improved canonical theory*.

It has been shown that canonical theory is equivalent to placing the transition state at
a Gibbs free energy maximum, in contrast to placing it at the col.

3.4 Quantum Transition State Theory

Quantum transition state theory puts classical theory into quantum terms, for instance
classical partition functions are replaced by quantum versions, the calculation of the
classical number of states of microcanonical theory becomes a quantum number of
states, and the transmission coefficient κ is now made to depend on either temperature
or energy in an attempt to correct for classical motion along the reaction coordinate.
The tunnelling corrections which must be made to classical transition state theory
become an integral part of a fully quantum treatment. Likewise, a fully quantum
treatment of motion along the reaction coordinate automatically removes the
separability of motion problems of classical theory.

3.4.1 Fully Quantum theories

These attempt to:

1. Calculate the quantum number of states at any point along the reaction coordinate.
2. Calculate the rate of movement along the reaction coordinate in a quantum manner.
    This replaces the classical internal translational rate of motion on the reaction
    coordinate, by a quantised rate, and this quantised rate does not assume that the
    internal translational motion must be separated from all the other motions of the
    (X..Y..Z) unit.

These are attempts at a totally quantum theory, but the calculations are extremely
difficult, and most approaches have concentrated on particular aspects of classical
transition state theory, and quantised these.

3.4.2 Variational quantum theory

One very successful approach is to formulate quantum analogues of microcanonical, canonical and improved canonical calculations. One version retains the approximation of separability of motion along the reaction coordinate, and therefore relies on a classical calculation of this motion.

Variational procedures are used to find the best dividing surface and the best generalised transition state. The calculations are similar to the classical version, only the number of quantum vibrational states is now calculated at each point along the reaction coordinate at which a dividing surface is placed. This is the number of quantum states with energies less than E which lie in the region of the potential energy surface cut by the dividing surface. The minimum number of states gives the generalised transition state at which there is a bottleneck to progress along the reaction coordinate. The rate of passage through the dividing surface, and the rate constant, both at constant total energy E, are then calculated *classically*.

Quantum canonical transition state theory looks for a dividing surface that minimises the thermal rate constant calculated from the "approximate" quantum microcanonical rate constant discussed above, and using a quantised distribution of energies. Improved quantum canonical theory suppresses the contribution to the rate constant from units with total energy below the threshold.

Attempts must now be made to correct for quantum effects on motion along the reaction coordinate, since in these approximate treatments this motion is treated classically. Quantisation shows that separability of motion along the reaction coordinate is invalid, and it also allows tunnelling to be possible. If classical motion along the reaction coordinate is used, as above, then tunnelling corrections must always be made.

Tunnelling corrections are simpler to make than corrections for "non separability", and several treatments have been proposed. One method of approximate solution is to consider trajectories on the "upside down" potential energy surface where values of the negative of the potential energy are plotted. This surface is the mirror image of the standard 3-dimensional potential energy surface, Figure 3.9. These trajectories are tunnelling paths for energies below the threshold. At an energy just below the threshold the trajectory goes through the col, but at lower energies the tunnelling path cuts the corner increasingly as the energy decreases.

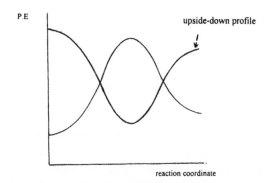

Figure 3.9 An "upside-down" potential energy profile for use in corrections for non-separability of motion along the reaction coordinate.

Corrections for tunnelling are even more important in variational transition state theory than in conventional theory. Any calculation which ignores tunnelling will underestimate the exact rate constants, and this is especially so at low temperatures. Variational minimisation displaces the barrier to reaction to energies lower than the col, and results in an even higher underestimate. One solution is to assign a transmission coefficient to each internal state; however, approximate values are generally used.

The reaction coordinate is the most likely path across the potential energy surface which also goes through the col. However, it is not the only path which can be used, and there are many possible paths all displaced from the reaction coordinate. The curvature of a path is a measure of its displacement from the reaction coordinate, Figure 3.10.

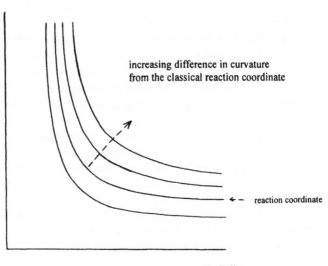

$$X + YZ \;\rightarrow\; XY + Z$$

Figure 3.10 Reaction paths showing increasing difference in curvature from the classical reaction coordinate.

Quantum calculations show that tunnelling is an inevitable feature of motion over the potential energy surface, and if corrections for tunnelling are to be made on classical motion across the surface, the path which gives the highest degree of tunnelling is the best. The shape of the barrier depends crucially on the curvature as well as determining the extent of tunnelling, and so the path giving the thinnest barrier should optimise the chance of tunnelling occurring.

Three paths are given in the following contour diagram:

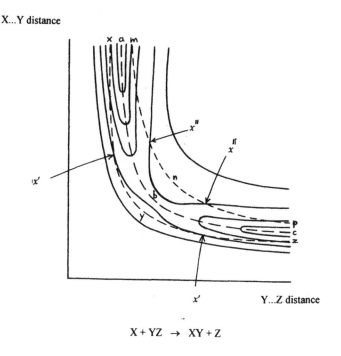

$$X + YZ \rightarrow XY + Z$$

Figure 3.11 A contour diagram showing three possible paths.

abc represents the classical reaction coordinate, mnp which is curved from the reaction coordinate gives the shortest route across the surface, and xyz displaced to the other side of the reaction coordinate gives the longest route of the three paths. Study of these paths shows that for the two outermost paths the distance to a given potential energy is shorter than the corresponding distance for the classical reaction coordinate. When it comes to the region of maximum potential energy, the distance between two equal potential energy contours is now shortest, $x'' - x''_c$, for path mnp, and longest, $x^1 - x^1$, for path xyz, with the classical reaction coordinate being intermediate. Furthermore, both these paths can reach higher potential energies than can the classical reaction coordinate. This is seen very easily on the corresponding potential energy profiles, Figure 3.12 (i), (ii), (iii).

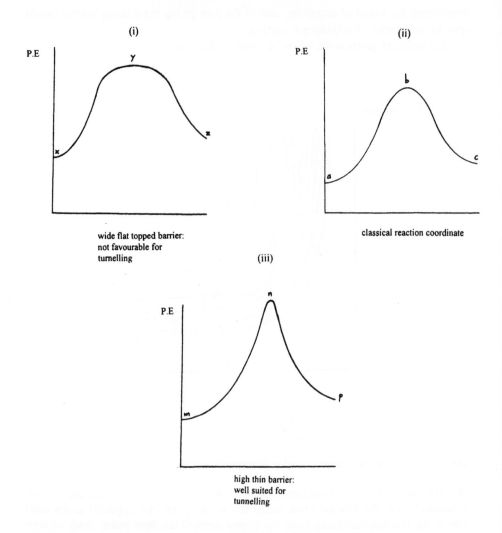

Figure 3.12 Potential energy profile for each of the paths given in Figure 3.11.

Figure (iii) gives a path which is well suited to tunnelling. However, there are many others which could also be well suited for tunnelling, and a decision has to be made as to which is the best path. Full calculations considering all possible paths are lengthy, and approximate methods are used.

Sometimes a parameter is chosen which depends on curvature as a way of specifying departure from the reaction coordinate. A variational procedure is carried out and allows the path to be chosen which minimises the rate, and so maximises the contributions from tunnelling.

A path which is often chosen as being nearest to the truth is one defined by the outermost turning points of the stretching vibration perpendicular to the reaction coordinate for the unit (X..Y..Z), Figure 3.13. This is a tunnelling path which is shorter than the classical path and which gives the thin, high barrier which promotes tunnelling, Figure 3.14.

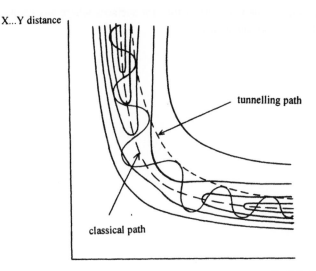

$$X + YZ \rightarrow XY + Z$$

Figure 3.13 A tunnelling path defined by the outermost turning points of the stretching vibration perpendicular to the reaction coordinate.

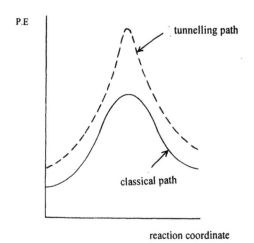

Figure 3.14 Potential energy profile showing the classical and tunnelling paths of Figure 3.13.

This has been shown to be very satisfactory for surfaces where any displacement from the classical reaction coordinate is small.

4

Basic Unimolecular Theory

Chapter 1 Section 1.3 gives a master mechanism for uni, bi and termolecular elementary reactions, and Table 1.2 gives a kinetic analysis of the three mechanisms. The mechanism of interest to this chapter is

$$A + A \quad \overset{k_1}{\rightarrow} \quad A^* + A \qquad \text{activation}$$

$$A^* + A \quad \overset{k_{-1}}{\rightarrow} \quad A + A \qquad \text{deactivation}$$

$$A^* \quad \overset{k_2}{\rightarrow} \quad \text{products} \qquad \text{reaction}$$

The crucial step in the development of unimolecular theory was the postulate of a time lag between activation and reaction, during which time an activated molecule can either be deactivated in an energy-transferring collision or it can alter configuration to reach the critical configuration and react. Table 1.2 discusses conditions in which each of these steps can be rate-determining, and where the rates of these steps are comparable.

The Marcus theory (Section 5.1) and extensions of it (Section 5.2), is the theory now currently used. But many concepts from earlier theories are implicitly incorporated into Marcus' theory, and these aspects will be discussed first. Unimolecular theory demonstrates clearly that each new development requires an increasingly more precise and detailed description of the activated molecule, the process of activation and the process of reaction.

4.1 Experimental observations and manipulation of experimental data

Within any given experiment, unimolecular reactions are strictly first order irrespective of the initial concentration or pressure. At high pressures the observed first order rate constant is strictly independent of pressure, but if experiments are carried out at low pressures or at intermediate pressures then the observed first order rate constants are found to depend on pressure.

Table 1.2 shows that, in general, for a unimolecular reaction

$$k_{obs}^{1st} = \frac{k_1 k_2 [A]}{k_{-1}[A] + k_2}. \tag{4.1}$$

At high pressures where k_{obs}^{1st} is strictly constant and independent of pressure, the equation becomes

$$\left(k_{obs}^{1st} \right)_{high\ pressures} = k_\infty = \frac{k_1 k_2}{k_{-1}}.$$

(4.2)

At low pressures k_{obs}^{1st} becomes dependent on pressure, so that

$$k_{obs}^{1st} = k_1 [A]$$

(4.3)

Values of k_1, k_{-1} and k_2 can be found from experimental data as follows:

1. When the rate of deactivation equals the rate of reaction

$$k_2 [A^*] = k_{-1} [A^*][A]$$

(4.4)

$$\therefore \qquad k_2 = k_{-1}[A]$$

(4.5)

then

$$k_{obs}^{1st} = \frac{k_1 k_2}{2k_{-1}}$$

(4.6)

$$= \frac{k_\infty}{2}.$$

(4.7)

Hence k_2 can be found from the pressure at which $k_{obs}^{1st} = k_\infty$ has fallen to $k_\infty/2$, provided the value of k_{-1} is known. This is generally taken to be a collision number, or the collision number Z modified by a collision efficiency λ.

$$k_2 = \lambda Z [A]_{\infty/2}.$$

(4.8)

Once k_2 is known, then k_1 can be found from k_∞

$$k_1 = \lambda Z \frac{k_\infty}{k_2}.$$

(4.9)

2. The experimental results can be given in the form of a graph of

$$\frac{1}{k_{obs}^{1st}} \qquad \text{against} \qquad \frac{1}{[A]}.$$

Rearrangement of Equation 4.1 gives

$$\frac{1}{k_{obs}^{1st}} = \frac{k_{-1}}{k_1 k_2} + \frac{1}{k_1 [A]}.$$

(4.10)

If k_1, k_{-1} and k_2 are constants then the graph should be linear with

$$\text{slope} = \frac{1}{k_1} \text{ and intercept} = \frac{k_{-1}}{k_1 k_2}$$

which provides an alternative route to k_2, again provided k_{-1} is known.

If, however, k_1 and k_2 are not single valued then the graph will be a curve. Experiments, in fact, demonstrate that graphs of $1/k_{obs}^{1st}$ against $1/[A]$ are decidedly curved, from which it follows that k_1 and k_2 cannot take unique values for any given reaction.

4.2 Significance and physical interpretation of the three rate constants in the general mechanism

1. The rate constant k_{-1}

In the theories to be discussed, k_{-1} takes a **unique** value and describes deactivation from **all** energies above the critical to **any** level below the critical. The magnitude is often taken to be λZ, where Z is the collision number, and λ the collision efficiency which lies between zero and unity.

2. The rate constant k_1

k_1 is the rate constant for activation. There are two models for activation, a simple one and a complex one. To make the differences between the models more obvious, these will be treated in parallel.

The simple model
In this model k_1 is unique for a particular reaction at a given temperature. There is, therefore, *one* k_1 describing activation to *all* levels above the critical energy ε_0, so there is no discrimination between these high energy levels. This assumption is used by Lindemann and Hinshelwood.

The complex model
In this model k_1 is not a constant but is allowed to take different values dependent on how much energy above the critical energy, ε_0, the activated molecule possesses. It is easier to produce an activated molecule with energy slightly above the critical energy, ε_0, than to produce an activated molecule with energy far in excess of ε_0. The k_1 for the first instance would thus be larger than k_1 for the second situation. This model is used in
1. Kassel's theory
2. Marcus' theory
3. Quack and Troe's theory
4. Slater's theory.

Once it has been postulated that a critical energy is required to define an activated molecule, it becomes pertinent to ask "how is this activation energy accumulated, and to what does it correspond physically?" Again there are two basic models.

The simple model

1. An activated molecule, A*, is one which has accumulated the critical energy ε_0 in *2* squared terms. For the *high* pressure region, statistical mechanics can calculate the fraction of molecules which can have an energy at least ε_0 in *2* squared terms, and this *fraction is equal to*

$$\exp\left(-\frac{\varepsilon_0}{kT}\right).$$

2. *Energy in 2 squared terms could correspond to*

(i) kinetic energy of relative translational motion along the line of centres of the two colliding molecules involved in an energy transforming collision, giving a contribution to the critical energy of $\frac{1}{2}mv^2$ from each molecule

(ii) rotational energy with two rotational modes each contributing *1* squared term

(iii) vibrational energy with one normal mode contributing *2* squared terms.

The complex model

1. An activated molecule, A*, is one which has accumulated the critical energy, ε_0, in *2s* squared terms, the value of *s* being unspecified. For the *high* pressure region, statistical mechanics can calculate the fraction of molecules which can have an energy at least ε_0 in *2s* squared terms, and *this fraction is approximately equal to*

$$\frac{(\varepsilon_0/kT)^{s-1}}{(s-1)!}\exp\left(-\frac{\varepsilon_0}{kT}\right).$$

2. *Energy in 2s squared terms could correspond to any combination of* relative translational motion, rotational motion and vibration, such as would add up to *2s* squared terms. If *s* is large, then vibrational modes will inevitably make the major contribution.

For unimolecular reactions, it is generally assumed that vibrational energy is what contributes to the activation process, and so

For the simplest model, activation energy at least ε_0 is accumulated in *one* vibrational mode or oscillator, that is in *2* squared terms. This criterion is relevant only to the most primitive unimolecular theory as postulated by Lindemann.

For the more complex model, energy at least ε_0 is accumulated in *2s* squared terms. This criterion is relevant to

1) Hinshelwood's theory
2) Kassel's theory
3) Marcus' theory
4) Quack and Troe's theory
5) Slater's theory.

3. *The rate constant k_2*

k_2 is a rate constant for reaction. This is related to the time lag which is the mean time interval between the activating collision and the moment of reaction, and is thus the mean lifetime of an A* molecule.

For a first order process

$$k = \frac{1}{\tau} \qquad (4.11)$$

where τ is the mean life time.

There are two assumptions which can be made about τ, and hence about k_2 in unimolecular theories, and again they can be classified as simple and complex.

The simple assumption
The mean time lag is *independent* of how much energy the activated molecule A* has above the critical energy, ε_0. This means that, *on average*, a very highly activated molecule lasts as long as one with energy only just above the critical energy, ε_0, before reaction occurs. And so there is *one k_2* which describes reaction from *all* energy levels above the critical energy ε_0.

A unique value of τ independent of energy is used in
1) the Lindemann treatment
2) Hinshelwood's theory.

The complex assumption
The mean time lag *depends* on how much energy the activated molecule A* has above the critical energy ε_0. Highly activated molecules will last, *on average*, for a much shorter length of time than would an A* molecule which has energy just above the critical energy ε_0. And so, there are a *series of k_2* values which describe reaction from *all* the various energy levels above the critical energy ε_0.

A series of τ values and a series of k_2 values are used in
1) Kassel's theory
2) Marcus' theory
3) Quack and Troe's theory
4) Slater's theory.

If there is a time lag between activation and reaction, it becomes pertinent to ask, "what happens to an activated molecule A* during the time lag?"

This is a very important question. The activated molecule has to alter configuration until it reaches the critical configuration, and can react. The physical processes involving possible configurational changes and possible energy redistributions are specific to each theory, and will be discussed in detail when each theory is dealt with in turn.

4.3 Aims of any theory of unimolecular reactions
Any theory of unimolecular reactions must include the following points.
1.   A model for the activation process must be given, and a value for k_1 then calculated. A single value for k_1 implies that no discrimination of the energy levels above the critical is being made, while a set of values depending on energy, $k_1(\varepsilon')$, where $\infty > \varepsilon' > \varepsilon_0$, implies that the rate constant for activation depends on the energy level to which activation is occurring.
2.   A model for the reaction process must be given, and a value of k_2 then calculated. As in the activation process, a unique value of k_2 implies that no discrimination of the energy levels from which reaction is occurring is being made, while a set of rate constants depending on energy, $k_2(\varepsilon')$, where $\infty > \varepsilon' > \varepsilon_0$, implies that the rate of reaction depends on how much energy in excess of ε_0 the activated molecule possesses.
3.   A value for the rate constant for deactivation, k_{-1} must be found. The rate of deactivation is taken to be independent of the energy level from which deactivation

occurs, and so k_{-1} takes a unique value. This model of deactivation is assumed in all the theories to be discussed, and k_{-1} is simply the collision number for the decomposing molecule. There is, however, no need to restrict deactivation in this way. It would be possible to consider that the rate of deactivation from high energy activated molecules A^* might be dependent on the energy above the critical energy, ε_0, which A^* possesses, that is, a collision with a high energy A^* would be more likely to remove energy than a collision with a low energy A^*. The calculation of k_{-1} could be then modified to account for this, giving a value of k_{-1} (ε').

4.   Once k_1 and k_{-1} and k_2 have been calculated the value of k_∞ can be predicted

$$k_\infty = \frac{k_1 k_2}{k_{-1}}.$$

(4.12)

5.   Any theory must be able to reproduce the experimental variation of k_{obs}^{1st} with pressure. This is by far the most difficult aspect of the theoretical treatment, and, for all the theories to be discussed, this section is easily the weakest part.

6.   Once the theoretical development is complete, the theory is tested against experiment, to see whether the model needs to be modified or approximations removed.

4.4 Aspects of vibrations used in unimolecular theory

The vibrations in a molecule can be regarded as harmonic or anharmonic oscillators.

4.4.1 Harmonic vibrations

If the vibrations are harmonic, the overall movements of the atoms in the molecule can be broken down into normal modes of vibration, each of which can be regarded as a harmonic oscillator and talked about as vibrational degrees of freedom. Each oscillator is made up of kinetic energy and potential energy, and the vibrational energy can be referred to as energy in 2 squared terms.

If the vibrations are regarded as harmonic, then there cannot be any flow of energy between the normal modes of vibration. Hence the distribution of energies between the normal modes of vibration must remain fixed, and can only be altered by a collision, or by interaction with radiation.

4.4.2 Anharmonic vibrations

If vibrations are anharmonic, the overall motion of the atoms in the molecule can only approximately be broken down into normal modes. When vibrations are anharmonic, energy is allowed to flow between the normal modes of vibration, and so the distribution of energy between the normal modes of vibration can alter between collisions. Vibrations are generally found to be anharmonic, though many theoretical calculations assume harmonicity, simply because of the mathematical complexity of allowing for possible flows of energy around the molecule.

4.4.3 Energies in oscillators, and energy in bonds

It is vital to distinguish between energy in an oscillator, and energy in a bond.

More than one oscillator, or normal mode of vibration, may contribute to the overall motion of the atoms in a bond, or only some may contribute, depending on the structure of the molecule. The oscillator *must not* be taken to mean an interatomic bond, or to represent bonds. Likewise the requirement of energy at least ε_0 in one oscillator must not be confused with energy in a bond, and a flow of energy into an oscillator is not the same as energy flowing into a bond. These distinctions must be appreciated before a detailed discussion of unimolecular theories is given.

Also when energy in s oscillators, s normal modes of vibration or in $2s$ squared terms is discussed, it must be remembered that discussion is centred around that part of the total energy which is associated with the $2s$ squared terms, and is not a discussion about the total energy.

4.5 Mathematical tools used in unimolecular theory
An outline of the mathematical tools common to all the theories is given below.

4.5.1 Equilibrium statistical mechanics
The use of equilibrium statistical mechanics is limited to the high pressure region where the equilibrium

$$A + A \underset{k_{-1}}{\overset{k_1}{\rightleftharpoons}} A^* + A$$

is likely to be set up. Equilibrium statistical mechanics must not be used for the low pressure region.

Before outlining the use of the Maxwell-Boltzmann distribution in unimolecular theory, several points should be made.

4.5.2 Calculation of k_1/k_{-1}
This uses the Maxwell-Boltzmann distribution to calculate the quantity

$$\frac{N_{\varepsilon' \to \varepsilon' + \delta\varepsilon}}{N_{\text{total}}} \tag{4.13}$$

which is the equilibrium ratio for the molecules with energy ε' lying in the short range $\delta\varepsilon$, where $\infty > \varepsilon' > \varepsilon_0$. This can also be expressed as a fraction, where the numerator is the number of activated molecules which have energy ε', which is at least ε_0, in the short range $\delta\varepsilon$ where $\infty > \varepsilon' > \varepsilon_0$. The denominator is the total number of all molecules present, activated and otherwise. Statement 4.13 is in effect a calculation of the ratio

$$\frac{[A^*]_{\varepsilon' \to \varepsilon' + \delta\varepsilon}}{([A^*] + [A])_{\text{total}}}$$

so that

$$\frac{[A^*]_{\varepsilon' \to \varepsilon' + \delta\varepsilon}}{([A^*] + [A])_{\text{total}}} = \frac{N_{\varepsilon' \to \varepsilon' + \delta\varepsilon}}{N_{\text{total}}}. \tag{4.14}$$

But what is required is the ratio of rate constants, k_1/k_{-1} again in this short range $\varepsilon' \to \varepsilon' + \delta\varepsilon$. This ratio is

$$\left(\frac{k_1}{k_{-1}}\right)_{\varepsilon' \to \varepsilon' + \delta\varepsilon} = \frac{[A^*]_{\varepsilon' \to \varepsilon' + \delta\varepsilon}}{[A]}. \tag{4.15}$$

If the approximation is made that $[A^*] \ll [A]$ then

$$\frac{[A^*]_{\varepsilon' \to \varepsilon' + \delta\varepsilon}}{[A]} = \frac{N_{\varepsilon' \to \varepsilon' + \delta\varepsilon}}{N_{total}}. \tag{4.16}$$

from which $(k_1/k_{-1})_{\varepsilon' \to \varepsilon' + \delta\varepsilon}$ can be found.

However, what is actually required finally in the calculations is an expression for k_{obs}^{1st} and for k_∞.

Equation 4.1 can be rearranged to the form

$$k_{obs}^{1st} = \frac{k_2 k_1 / k_{-1}}{1 + k_2 / k_{-1}[A]}. \tag{4.17}$$

The procedure to be followed in obtaining an expression for k_{obs}^{1st} depends on the model assumed for the reaction step.

1. If all activated molecules last, on average, for the same length of time then k_2, the rate constant for the reaction step, will be a constant. If k_{-1} is also assumed to be a constant, and equal to the collision number for the reacting molecule then equation 4.17 becomes

$$\left(k_{obs}^{1st}\right)_{\varepsilon' \to \varepsilon' + \delta\varepsilon} = \left(\frac{k_2}{1 + k_2 / k_{-1}[A]}\right)\left(\frac{k_1}{k_{-1}}\right)_{\varepsilon' \to \varepsilon' + \delta\varepsilon} = \text{a constant}\left(\frac{k_1}{k_{-1}}\right)_{\varepsilon' \to \varepsilon' + \delta\varepsilon} \tag{4.18}$$

and so when $(k_{obs}^{1st})_{\varepsilon' \to \varepsilon' + \delta\varepsilon}$ is summed over all energies from $\varepsilon_0 \to \infty$ it is only the term $(k_1/k_{-1})_{\varepsilon' \to \varepsilon' + \delta\varepsilon}$ which will be summed.

An explicit formulation of k_2 will not be needed before summation occurs. This argument is applicable to the Lindemann treatment and the Hinshelwood theory.

2. If activated molecules have lifetimes which depend on the amount of energy which they possess in excess of the critical energy ε_0, then k_2 will no longer be a constant, but will depend on the value of ε'. Under these conditions

$$\left(k_{obs}^{1st}\right)_{\varepsilon' \to \varepsilon' + \delta\varepsilon} = \frac{(k_2)_{\varepsilon' \to \varepsilon' + \delta\varepsilon}\left(\dfrac{k_1}{k_{-1}}\right)_{\varepsilon' \to \varepsilon' + \delta\varepsilon}}{1 + \dfrac{(k_2)_{\varepsilon' \to \varepsilon' + \delta\varepsilon}}{k_{-1}[A]}} \tag{4.19}$$

and this must be totally formulated before summation is carried out.

An explicit formulation of the dependence of k_2 on energy must be found before the summation can be carried out. This argument is applicable to the theories of Kassel, Marcus and its extensions by Troe.

4.5.3 Summations and integrals

When *classical* theories are used this means that the energy is continuously variable. This means that in summing an expression over a range in energy, the summations can be replaced by an integral. In *quantum* theories a direct summation must be made over all quantum states.

4.5.4 Probability theory

In Kassel's theory probability theory is used. This is because the equation requires calculation of quantities like, "the chance that one oscillator has an energy at least ε_0, while all of a total of s oscillators have between them an energy, ε' ".

4.5.5 The actual Maxwell-Boltzmann calculations

The quantity , $N_{\varepsilon' \to \varepsilon' + \delta\varepsilon}/N_{total}$, is the fraction of molecules which have energy ε' in the short range $\delta\varepsilon$ where $\infty > \varepsilon' > \varepsilon_0$, and this can give the fraction of molecules with energy at least ε_0 in 2 squared terms, or, more generally, in $2s$ squared terms. When the energy is in $2s$ squared terms, that is in s oscillators, there need be no explicit statement as to how the energy is distributed among the $2s$ squared terms.

   The results of the calculations are:

(i) The fraction of molecules which have energy ε' which is at least ε_0 in 2 squared terms, and where ε' lies in the short range $\delta\varepsilon$ is

$$\frac{N_{\varepsilon' \to \varepsilon' + \delta\varepsilon}}{N_{total}} = \frac{1}{kT} \exp\left(-\frac{\varepsilon'}{kT}\right) \delta\varepsilon.$$

(4.20)

(ii) The fraction of molecules which have energy ε' which is at least ε_0 in $2s$ squared terms, and where ε' lies in the short range $\delta\varepsilon$ is

$$\frac{N_{\varepsilon' \to \varepsilon' + \delta\varepsilon}}{N_{total}} = \frac{1}{kT(s-1)!} \left(\frac{\varepsilon'}{kT}\right)^{s-1} \exp\left(-\frac{\varepsilon'}{kT}\right) \delta\varepsilon.$$

(4.21)

These expressions must now be summed over all energies from ε_0 to ∞.

(a) For case (i) immediately above the summation becomes

$$\sum_{\varepsilon' = \varepsilon_0}^{\infty} \frac{N_{\varepsilon' \to \varepsilon' + \delta\varepsilon}}{N_{total}} = \sum_{\varepsilon' = \varepsilon_0}^{\infty} \frac{1}{kT} \exp\left(-\frac{\varepsilon'}{kT}\right) \delta\varepsilon.$$

(4.22)

Making the approximation that $[A^*] \ll [A]$ given in equation 4.16, then

$$\frac{k_1}{k_{-1}} = \frac{[A^*]}{[A]} = \sum_{\varepsilon' = \varepsilon_0}^{\infty} \frac{1}{kT} \exp\left(-\frac{\varepsilon'}{kT}\right) \delta\varepsilon.$$

(4.23)

The summation must now be carried out. A direct summation over all quantum states is required in the quantum form. Provided the spacings of the levels are not too large the quantum expression after direct summation can be taken to be approximately equal to the classical form.

In the classical form the summation is replaced by an integral between the same limits ε_0 and ∞.

$$\frac{k_1}{k_{-1}} = \frac{[A^*]}{[A]} = \int_{\varepsilon_0}^{\infty} \frac{1}{kT} \exp\left(-\frac{\varepsilon'}{kT}\right) d\varepsilon \tag{4.24}$$

$$= \exp\left(-\frac{\varepsilon_0}{kT}\right). \tag{4.25}$$

(b) For case (ii) above the summation becomes

$$\sum_{\varepsilon'=\varepsilon_0}^{\infty} \frac{N_{\varepsilon'\to\varepsilon'+\delta\varepsilon}}{N_{\text{total}}} = \sum_{\varepsilon'=\varepsilon_0}^{\infty} \frac{1}{kT(s-1)!}\left(\frac{\varepsilon'}{kT}\right)^{s-1} \exp\left(-\frac{\varepsilon'}{kT}\right)\delta\varepsilon. \tag{4.26}$$

Again making the assumption that $[A^*] \ll [A]$ then

$$\frac{k_1}{k_{-1}} = \frac{[A^*]}{[A]} = \sum_{\varepsilon'=\varepsilon_0}^{\infty} \frac{1}{kT(s-1)!}\left(\frac{\varepsilon'}{kT}\right)^{s-1} \exp\left(-\frac{\varepsilon'}{kT}\right)^{s-1}\delta\varepsilon. \tag{4.27}$$

As in the simple case, direction summation over all quantum states is required for the quantum version, but integration can be carried out for a classical version.

$$\frac{k_1}{k_{-1}} = \frac{[A^*]}{[A]} = \int_{\varepsilon_0}^{\infty} \frac{1}{kT(s-1)!}\left(\frac{\varepsilon'}{kT}\right)^{s-1} \exp\left(-\frac{\varepsilon'}{kT}\right) d\varepsilon. \tag{4.28}$$

If $\varepsilon_0 \gg (s-1)kT$, then

$$\frac{k_1}{k_{-1}} = \frac{[A^*]}{[A]} \approx \frac{1}{(s-1)!}\left(\frac{\varepsilon_0}{kT}\right)^{s-1} \exp\left(-\frac{\varepsilon_0}{kT}\right). \tag{4.29}$$

4.6 Unimolecular theories
Several unimolecular theories have been proposed. They are

1. Lindemann's treatment
2. Hinshelwood's theory
3. Kassel's theory
4. Rice, Ramsperger, Kassel and Marcus theory
5. Quack and Troe theory (Adiabatic Channel Theory)
6. Slater's theory.

All these theories have a model describing activation and reaction. Taken chronologically these models become more detailed, precise and complex. Each theory will be taken in turn, starting with

(i)   a general statement of the problems involved,
(ii)   the models for activation and reaction,
(iii)   the calculation,
(iv)   comparison of theory with experiment.

4.7 The simple Lindemann treatment
This is a very primitive treatment using the simplest model.

4.7.1 Criterion of activation
Energy at least ε_0 must be accumulated in 2 squared terms before a molecule can be activated. The criterion is general, with no enquiry being made as to what this energy corresponds.

4.7.2 Criterion of reaction
A molecule with an energy ε_0 will react, but there is no enquiry into how the activated molecule A^* gets into the critical configuration.

4.7.3 The Lindemann calculation
The rate of collision modified to take account of the fact that not all molecules react is calculated. Molecules which have energy in excess of the critical can react, and if this energy is in 2 squared terms the fraction of molecules having this energy is given by Equation 4.20 so that

$$\frac{k_1}{k_{-1}} = \exp\left(-\frac{\varepsilon_0}{kT}\right).$$

$$(4.30)$$

If k_{-1} is a constant and equal to a collision number Z, then

$$k_1 = Z \exp\left(-\frac{\varepsilon_0}{kT}\right)$$

$$(4.31)$$

and has the physical significance that, no matter how much energy in excess of ε_0 the activated molecule possesses, one collision will be sufficient to remove this energy.

The value of k_2, which is the rate constant for reaction, can be found from the value of k_1/k_{-1} and the high pressure first order rate constant k_∞.

$$k_\infty = \left(k_{obs}^{1st}\right) = \frac{k_1 k_2}{k_{-1}}.$$

$$(4.32)$$

If k_1, k_{-1} and k_2 are constants then the graph of

$$\frac{1}{k_{obs}^{1st}} \text{ v } \frac{1}{[A]}$$

should be linear with

$$\text{slope} = \frac{1}{k_1} \text{ and intercept} = \frac{k_{-1}}{k_1 k_2}$$

and this should give an alternative route to k_2.

4.7.4 Comparison of theory and experiment for the Lindemann treatment
(i)   A comparison of the theoretical k_1 with the observed k_1 shows that

$$k_1 \text{ (observed)} \quad >> \quad k_1 \text{ (calculated)}$$

implying that accumulation of energy cannot be as described. Energy must be in more than two squared terms, since this is the only way in which k_1 could be increased in the drastic way needed to bring agreement with experiment.

(ii) A comparison of the pressure at which a falling off in the observed first order rate constant becomes apparent, shows that the observed values are much lower than those predicted. The decreasing value of the observed rate constant as the pressure is lowered is a result of the failure of collisions to maintain the adequate supply of A* found at high pressures. However, if inert gases are added to increase the collision rate, then the observed first order rate constant can be brought up to its high pressure value. This is in keeping with prediction.

(iii) The graph of

$$\frac{1}{k_{obs}^{1st}} \text{ v } \frac{1}{[A]}$$

is quite decidedly curved, in direct contradiction of the linear graph which should result if k_2 takes a single unique value.

These experimental observations imply that the simple Lindemann model must be modified to allow a more efficient activation process, and to allow k_1 and k_2 to take different values dependent on the energy level above the critical *to which* activation is taking place, and *from which* reaction is occurring.

Hinshelwood's theory incorporates the more efficient activation process, but maintains a unique k_2. Kassel, Marcus and Slater consider the more efficient activation process and allow k_2 to take values which depend on energy.

4.8 Hinshelwood's Theory
Energy can now be accumulated in *2s* squared terms so that internal modes can be utilised.

Details of how the internal modes are involved, specification of how many modes are utilised and the process of distribution of the energy into the normal modes are not discussed. But if the number of modes is large then the energy must be accumulated predominantly in vibration. The theory does not discuss how the activated molecule moves into an activated complex, it merely states that once the necessary energy has been accumulated, then the critical configuration *can* be attained.

Unique values for k_1 and k_2 are assumed, hence both are independent of how much energy greater than the critical energy, ε_0, the molecule has. These high energies are not discriminated.

4.8.1 Criterion of activation

A molecule is activated if it has an energy at least ε_0 in $2s$ squared terms. The value of s is unspecified, and identification of the modes involved is not made.

4.82 Criterion of reaction

The activated molecule lasts until, by an internal distribution of energy among the s normal modes, the relevant geometrical arrangement of the various atoms in the molecule is that corresponding to the critical configuration. This is a very vague statement, and is a consequence of the fact that the theory does not enquire into how this internal redistribution occurs, or how the critical configuration is attained.

4.8.3 The Hinshelwood calculation

Since energy is accumulated in s oscillators, or $2s$ squared terms, then the appropriate Maxwell-Boltzmann equations for $2s$ squared terms must be used. The statement (4.13) is the starting point from which Equations 4.21 and 4.26 to 4.29 follow. Summation of $(k_1/k_{-1})_{\varepsilon \to \varepsilon + \delta\varepsilon}$ can be carried out independently of the rate constant for the reaction step. This is because k_2 is a constant and does not depend upon the energy above the critical from which any given activated molecule reacts. Equation 4.29 gives the expression for k_1/k_{-1}, and if the collision number Z is used for k_{-1}, then Equation 4.33 follows.

$$k_1 = \frac{Z}{(s-1)!}\left(\frac{\varepsilon_0}{kT}\right)^{s-1} \exp\left(-\frac{\varepsilon_0}{kT}\right).$$

(4.33)

The next step is a calculation of the dependence of k_{obs}^{1st} as a function of pressure. Equation 4.17 taken in conjunction with 4.33 gives

$$k_{obs}^{1st} = \frac{k_2}{1+k_2/k_{-1}[A]}\frac{1}{(s-1)!}\left(\frac{\varepsilon_0}{kT}\right)^{s-1} \exp\left(-\frac{\varepsilon_0}{kT}\right)$$

(4.34)

where k_2 is a constant. This equation requires explicit values of k_1, k_{-1}, k_2, ε_0 and s.

A first approximate value of k_2 can be found from the pressure at which the observed first order rate constant has fallen to half its constant high pressure value, (Section 4.1). At this pressure

$$k_2 = k_{-1}[A]$$

(4.35)

and if k_{-1} is identified with the collision number for the reactant molecule then

$$k_2 = \lambda Z[A] \ .$$

(4.36)

Once this first approximate value of k_2 is found, a first approximate value for k_1 can be found as

$$k_1 = \lambda Z \frac{k_\infty}{k_2}. \tag{4.37}$$

Thereafter, determination of k_2, ε_0 and s is made from a best fit of experiment with theory, and a best fit k_1 value is then found by substitution of the best fit values for ε_0 and s into Equation 4.33.

Equation 4.2 allows k_∞ to be calculated

$$k_\infty = \frac{k_1 k_2}{k_{-1}}. \tag{4.38}$$

$$k_\infty = k_2 \frac{1}{(s-1)!} \left(\frac{\varepsilon_0}{kT} \right)^{s-1} \exp\left(-\frac{\varepsilon_0}{kT} \right) \tag{4.39}$$

where best fit values for k_2, ε_0 and s are again used.

4.8.4 Results from the Hinshelwood treatment

(i)   For a given T and ε_0, k_1 in the Hinshelwood expression (4.33) increases as s increases, which is what is required. It is found that the best fit with experiment is often obtained when about one half of the internal modes are utilised.

(ii) In the simple Lindemann treatment Equations 4.2 and 4.25 give

$$k_\infty = k_2 \exp\left(-\frac{\varepsilon_0}{kT} \right) \tag{4.40}$$

where k_2 is a constant. Because of this, strict Arrhenius behaviour is predicted in which the graph of log k_∞ against $1/T$ is linear.

However in the Hinshelwood treatment k_1/k_{-1} is now given by Equation 4.29 and so

$$k_\infty = \frac{k_2}{(s-1)!} \left(\frac{\varepsilon_0}{kT} \right)^{s-1} \exp\left(-\frac{\varepsilon_0}{kT} \right). \tag{4.41}$$

The pre-exponential term has in it the term $(\varepsilon_0/kT)^{s-1}$, and this will vary with temperature, so that a curve now is predicted for the graph of log k_∞ against $1/T$. Non-Arrhenius behaviour is predicted, but linear plots are found.

4.8.5 Comparison of theory with experiment

(i)   Allowing activation into s internal modes gives a calculated k_1 more in line with experiment, and results in a considerable improvement in predicting the pressure at which a fall off in the values of k_{obs}^{1st} is found.

(ii)   Best fit values of k_1, k_2 ε_0 and s are found. When curve fitting techniques are used there may be more than one set of k_1, k_2, ε_0 and s which could give agreement, and a final choice must be made from these.

(iii) The major criticism comes on two counts.

(a) Hinshelwood's theory predicts a curved Arrhenius plot, whereas experimental data give linear plots. But, if s is small, then the predicted curvature would be sufficiently slight so that only highly accurate data could detect the expected curvature.

(b) The most damaging criticism is that the theory does not account for the very decided curvature observed in the plots of

$$\frac{1}{k_{obs}^{1st}} \quad \text{against} \quad \frac{1}{[A]}.$$

The prediction of linearity is the most serious defect of the theory.

4.8.6 Areas of the Hinshelwood theory which need to be improved
The activation step
It is physically more realistic to think of a series of energies all lying above ε_0, with separate independent values of k_1, each describing activation to each different energy. Activation to a high energy above the critical would be more difficult (a small k_1) than activation to a level *just* above the critical (a larger k_1).

It is also more realistic to think of a series of energies all lying below the critical energy ε_0, rather than an indiscriminate situation of undifferentiated levels below the critical. This would require a series of k_1 values describing activation from specific energies below the critical to specific energies above the critical. The magnitude of k_1 would then depend on the *change in energy* involved, with small energy jumps having larger k_1 values, and larger energy jumps having smaller k_1 values. This possibility has *not* been considered in the theories which will be discussed.

The reaction step
Hinshelwood assumes that k_2 is a constant which is independent of the value of the energy above the critical energy ε_0 from which reaction is taking place. This then assumes that a very highly activated molecule A* will last as long as an A* molecule with energy only just above the critical. However, it is more likely that highly activated molecules have shorter lifetimes than those which are just activated. And so, a series of k_2 values should be assumed, with the magnitude of each k_2 depending on the energy value from which reaction is taking place.

4.8.7 A modified mechanism and calculation

$$A + A \quad \xrightarrow{k_1(\varepsilon')} \quad A^* + A$$

$$A^* + A \quad \xrightarrow{k_{-1}} \quad A + A$$

$$A^* \quad \xrightarrow{k_2(\varepsilon')} \quad \text{products}$$

where $k_1(\varepsilon')$ represents a whole series of k_1 values

$k_2(\varepsilon')$ represents a whole series of k_2 values

but k_{-1} is still taken to be a constant equal to the collision number.

The calculation then becomes

$$\text{Total rate of reaction} = \sum_{\varepsilon'} \text{rate of reaction from any particular } \varepsilon' \quad (4.42)$$

$$= \sum_{\varepsilon'} k_2(\varepsilon')[A^*] \tag{4.43}$$

$$= \sum_{\varepsilon'} \frac{k_1(\varepsilon')k_2(\varepsilon')[A]^2}{k_{-1}[A] + k_2(\varepsilon')} \tag{4.44}$$

$$k_{obs}^{1st} = \sum_{\varepsilon'} \frac{k_1(\varepsilon')k_2(\varepsilon')[A]}{k_{-1}[A] + k_2(\varepsilon')} \tag{4.45}$$

$$k_{\infty} = \sum_{\varepsilon'} \frac{k_1(\varepsilon')k_2(\varepsilon')}{k_{-1}}. \tag{4.46}$$

To be able to proceed further requires being more explicit about the physical details of the model assumed. Doing this leads naturally to the theories of Kassel and Marcus, and Marcus extensions.

4.9 Kassel's Theory

The Hinshelwood theory was primarily interested in considering the effect of allowing internal modes to contribute to the accumulation of activation energy, and did not consider what happened to the activated molecule when it reacted. The Kassel theory is very much concerned with this process of getting from the activated molecule to the activated complex. Energy is allowed to flow around the molecule until conditions are just right for reaction. Kassel's original theory was a classical theory which was later put in quantum terms, and this quantum version is vastly superior to the classical version for fitting experiment with theory. The classical theory is the one which has generally been used, and which will, therefore, be developed here, but it will be pointed out where changes are needed to convert it to a quantum version.

4.9.1 The criterion for activation
(a) Vibrational energy is involved in activation.
(b) All the normal modes of vibration are assumed to be harmonic oscillators and have the same frequency.
(c) For a molecule to be activated, it must have energy, ε', and of that energy ε' an amount which is at least ε_0 must be found in s oscillators, that is in $2s$ squared terms. ε' can be specified by the summary statement $\infty > \varepsilon' \geq \varepsilon_0$.

4.9.2 The criterion for reaction
(a) Before the activated molecule can react there must be a *flow* of *energy* at least ε_0 into *one critical harmonic oscillator*. Once this has happened the activated molecule will have the correct distribution of vibrational energy for it to be able to react.
(b) The time taken for this redistribution of energy to occur is the time lag.

4.9.3 The Kassel mechanism

$$A + A \quad \xrightarrow{k_1(\varepsilon')} \quad A^* + A$$

$$A^* + A \quad \xrightarrow{k_{-1}(\varepsilon')} \quad A + A$$

$$A^* \quad \xrightarrow{k_2(\varepsilon')} \quad A^{\otimes}$$

$$A^{\otimes} \quad \xrightarrow{k_3} \quad \text{products}$$

(a) In this mechanism A^* is an activated molecule which has energy ε', of which an amount at least ε_0 is in s harmonic oscillators.
(b) $A^{\otimes}$ is the special activated molecule which must be formed before the critical configuration can be attained, and the $A^{\otimes}$ molecule must also have energy at least ε_0 in *one critical* harmonic oscillator.
(c) $k_1(\varepsilon')$ and $k_2(\varepsilon')$ represent a series of rate constants which depend on how much larger ε' is than ε_0.

4.9.4 The Kassel calculation
This calculation can be split up into stages, each dealing with a particular physical situation.

Step 1. *Calculation of* $(k_1/k_{-1})_{\varepsilon \,\to\, \varepsilon\,+\,\delta\varepsilon}$ *which is required for substitution into the equation which describes the dependence of* k_{obs}^{1st} *on pressure:*

$$k_{obs}^{1st} = \sum_{\text{all } \varepsilon'} \frac{k_2(\varepsilon')k_1(\varepsilon')/k_{-1}}{1 + k_2(\varepsilon')/k_{-1}[A]}. \tag{4.47}$$

This requires calculation of the fraction of molecules which have an energy ε' which is at least ε_0, and which has this energy in $2s$ squared terms in the short range $\delta\varepsilon$. This has already been calculated in Section 4.5.5 , Equation 4.21, and allows a calculation of k_1/k_{-1} for energies in the given short range $\varepsilon' \to \varepsilon' + \delta\varepsilon$, where $\infty > \varepsilon' \geq \varepsilon_0$

$$\left(\frac{k_1}{k_{-1}} \right)_{\varepsilon' \to \varepsilon' + \delta\varepsilon} = \frac{1}{kT(s-1)!} \left(\frac{\varepsilon'}{kT} \right)^{s-1} \exp\left(-\frac{\varepsilon'}{kT} \right) \delta\varepsilon \tag{4.48}$$

It is at this stage that the calculation diverges from that in the Hinshelwood treatment, where in Equation 4.18 k_2 is a constant, and it is only the term $(k_1/k_{-1})_{\varepsilon \to \varepsilon + \delta\varepsilon}$ which has to be summed.

But if k_2 is allowed to depend on the value of the energy ε' which is above the critical, as it does in the Kassel treatment, then Equation 4.19 must be used, and before this expression can be summed an explicit expression for the dependence of k_2 on energy must be formulated.

In the Kassel theory once a molecule has energy which is at least ε_0 in s harmonic oscillators, reaction *can* occur in due course, but will *not* occur *until* a flow of energy results in there being energy at least ε_0 *in one critical harmonic oscillator.*

Step 2. *The problem here is to rephrase the statement immediately above into mathematical terms, that is to obtain an expression for $k_2(\varepsilon')$*

What is required is a calculation of the probability that one oscillator has an energy at least ε_0, while all the total s oscillators have between them, on average, an energy ε', where $\infty > \varepsilon' \geq \varepsilon_0$.

This probability can be shown to be

$$W = \left(\frac{\varepsilon' - \varepsilon_0}{\varepsilon'} \right)^{s-1} \tag{4.49}$$

This probability has to be related to the concentration of A* and $A^\otimes$ so that W can be defined in terms of [A*] and $[A^\otimes]$.

The probability W is equal to the fraction of molecules which have energy at least ε_0 in one oscillator, that is $A^\otimes$ molecules, out of all the molecules which have energy ε' in s oscillators, that is A* molecules.

$$\frac{[A^\otimes]}{[A*]+[A^\otimes]} \approx \frac{[A^\otimes]}{[A*]} = \left(\frac{\varepsilon' - \varepsilon_0}{\varepsilon'} \right)^{s-1} \tag{4.50}$$

where the assumption that $[A^\otimes] \ll [A*]$ is made.

This result is required for a theoretical expression for $k_2(\varepsilon')$ and the link is the steady state expression which relates $k_2(\varepsilon')$ to $[A^\otimes]/[A*]$.

Step 3. *Steady state equations for [A*] and [A$^\circledcirc$]*

For any particular ε' the steady state treatment gives a set of equations relevant to that *one particular energy ε'*. If all possible energies ε' are then considered a whole series of sets of steady state equations will be needed for all ε', where ε' represents all energies equal to or greater than the critical energy ε_0.

One set of steady states using the Kassel mechanism given in Section 4.9.3

$$\frac{-d[A^*]}{dt} = k_1(\varepsilon')[A]^2 - k_{-1}[A^*][A] - k_2(\varepsilon')[A^*] = 0 \tag{4.51}$$

$$[A^*] = \frac{k_1(\varepsilon')[A]^2}{k_{-1}[A] + k_2(\varepsilon')}. \tag{4.52}$$

$$\frac{-d[A^\circledcirc]}{dt} = k_2(\varepsilon')[A^*] - k_3[A^\circledcirc] = 0 \tag{4.53}$$

$$[A^\circledcirc] = \frac{k_2(\varepsilon')[A^*]}{k_3}. \tag{4.54}$$

From Equation 4.54 and adding the subscript $\varepsilon' \to \varepsilon' + \delta\varepsilon$ to emphasise that these equations are for one value of ε' in a given short range $\varepsilon' \to \varepsilon' + \delta\varepsilon$, then

$$k_2(\varepsilon')_{\varepsilon' \to \varepsilon' + \delta\varepsilon} = k_3 \left(\frac{[A^\circledcirc]}{[A^*]} \right)_{\varepsilon' \to \varepsilon' + \delta\varepsilon}, \tag{4.55}$$

$$k_2(\varepsilon')_{\varepsilon' \to \varepsilon' + \delta\varepsilon} = k_3 \left(\frac{\varepsilon' - \varepsilon_0}{\varepsilon'} \right)^{s-1}_{\varepsilon' \to \varepsilon' + \delta\varepsilon}. \tag{4.56}$$

The rate for any given ε' in the short range $\varepsilon' \to \varepsilon' + \delta\varepsilon$

$$\text{Rate}(\varepsilon')_{\varepsilon' \to \varepsilon' + \delta\varepsilon} = k_3[A^\circledcirc]_{\varepsilon' \to \varepsilon' + \delta\varepsilon} \tag{4.57}$$

$$= k_2(\varepsilon')_{\varepsilon' \to \varepsilon' + \delta\varepsilon}[A^*] \tag{4.58}$$

$$= \frac{k_2(\varepsilon')_{\varepsilon' \to \varepsilon' + \delta\varepsilon}[A]\left(\dfrac{k_1}{k_{-1}}\right)_{\varepsilon' \to \varepsilon' + \delta\varepsilon}}{1 + \dfrac{k_2(\varepsilon')_{\varepsilon' \to \varepsilon' + \delta\varepsilon}}{k_{-1}[A]}} \tag{4.59}$$

This is the rate of reaction for $A^\circledcirc$ molecules lying in one short range of energies $\varepsilon' \to \varepsilon' + \delta\varepsilon$.

Step 4. *The rate for one energy ε' given in step 3 must now be summed to give the total rate and an expression for the dependence of k_{obs}^{1st} on pressure, and to give an expression for k_∞*

Expression 4.59 must now be summed over all ε'. The total rate $= k_{obs}^{1st}[A]$

$$= \sum_{\varepsilon'} \text{rate for each particular } \varepsilon' \text{ in the short range } \varepsilon' \to \varepsilon' + \delta\varepsilon \qquad (4.60)$$

$$= \frac{k_2(k_1/k_{-1})[A]}{1+k_2/k_{-1}[A]}. \qquad (4.61)$$

$$k_{obs}^{1st} = \sum_{\varepsilon'} \frac{k_2(\varepsilon')_{\varepsilon' \to \varepsilon' + \delta\varepsilon}\left(\dfrac{k_1}{k_{-1}}\right)_{\varepsilon' \to \varepsilon' + \delta\varepsilon}}{1 + \dfrac{k_2(\varepsilon')_{\varepsilon' \to \varepsilon' + \delta\varepsilon}}{k_{-1}[A]}} \qquad (4.62)$$

$$= \sum_{\varepsilon'} \frac{\dfrac{1}{kT(s-1)!}\left(\dfrac{\varepsilon'}{kT}\right)^{s-1}\exp\left(-\dfrac{\varepsilon_0}{kT}\right)k_3\left(\dfrac{\varepsilon'-\varepsilon_0}{\varepsilon'}\right)^{s-1}\delta\varepsilon}{1+k_3\left(\dfrac{\varepsilon'-\varepsilon_0}{\varepsilon'}\right)^{s-1}\Big/k_{-1}[A]}. \qquad (4.63)$$

Since the theory is a classical one, the energy is continuously variable, and so the summation can be replaced by an integral between the same limits, ε' ranges from ε_0 to ∞.

$$k_{obs}^{1st} = \frac{k_3}{kT(s-1)!}\int_{\varepsilon_0}^{\infty} \frac{\left(\dfrac{\varepsilon'}{kT}\right)^{s-1}\exp\left(-\dfrac{\varepsilon'}{kT}\right)\left(\dfrac{\varepsilon'-\varepsilon_0}{\varepsilon'}\right)^{s-1}d\varepsilon}{1+k_3\left(\dfrac{\varepsilon'-\varepsilon_0}{\varepsilon'}\right)^{s-1}\Big/k_{-1}[A]}. \qquad (4.64)$$

The final stage of the derivation is to formulate k_∞

$$k_\infty = \sum_{\varepsilon'} \frac{k_1(\varepsilon')_{\varepsilon' \to \varepsilon' + \delta\varepsilon}\, k_2(\varepsilon')_{\varepsilon' \to \varepsilon' + \delta\varepsilon}}{k_{-1}}\delta\varepsilon \qquad (4.65)$$

$$= \sum_{\varepsilon'} \frac{1}{kT(s-1)!}\left(\dfrac{\varepsilon'}{kT}\right)^{s-1}\exp\left(-\dfrac{\varepsilon'}{kT}\right)k_3\left(\dfrac{\varepsilon'-\varepsilon_0}{\varepsilon'}\right)^{s-1}\delta\varepsilon.$$

$$\qquad (4.66)$$

Again the sum can be replaced by an integral over the range $\varepsilon_0 \to \infty$.

$$\frac{k_3}{kT(s-1)!} \int_{\varepsilon_0}^{\infty} \left(\frac{\varepsilon'}{kT}\right)^{s-1} \exp\left(-\frac{\varepsilon'}{kT}\right)\left(\frac{\varepsilon'-\varepsilon_0}{\varepsilon'}\right)^{s-1} d\varepsilon \tag{4.67}$$

which eventually comes to

$$k_\infty = k_3 \exp\left(-\frac{\varepsilon_0}{kT}\right). \tag{4.68}$$

4.9.5 Comparison of theory with experiment
In contrast to Hinshelwood's theory, the Kassel theory predicts that a plot of

$$\frac{1}{k_{obs}^{1st}} \quad \text{against} \quad \frac{1}{[A]}$$

should be a curve, as is observed experimentally, and the corresponding calculated curve gives a relatively good fit to the experimental plot. This confirms that the assumption that the rate of decomposition of an activated molecule depends on how much energy the molecule has above the critical must be an essential feature of any theory.

The theory also predicts a linear Arrhenius plot, which is what is found experimentally.

Unfortunately there is no theoretical prediction of the value of s_f and s, ε_0 and k_3 are found by curve fitting to the fall off region, bringing with it all the ambiguities of curve-fitting.

The value of k_3 is often found to be around 10^{13} s^{-1}, but there are some reactions where k_3 is much larger, and no satisfactory explanation is available. The value of s is often around one half of the total normal modes of vibration leaving the question, "Why does energy in some modes not contribute to the activation energy?"

The problems really lie in the fact that classical theories simply are not valid. The classical version is the limiting form of the quantum version *if*, and *only if*, $h\nu \ll kT$ where ν is the fundamental vibration frequency for the critical oscillator.

4.10 Kassel's Quantum Theory
The criteria of activation and reaction have to be given in quantum terms, and a direct summation must be carried out.

4.10.1 Criterion of activation
1. Vibrational energy is involved in activation.
2. Each normal mode of vibration is assumed to be a harmonic oscillator, all having the same frequency.
3. The energies are now expressed in quanta, the critical number of quanta being $m = \varepsilon_0/h\nu$
4. For a molecule to be activated it must have n quanta, $n = \varepsilon'/h\nu$ where n is at least as great as m quanta, $m = \varepsilon_0/h\nu$ in s harmonic oscillators.

4.10.2 Criterion of reaction

Before a molecule can react there must be a flow of energy such that a number at least m quanta, $m = \varepsilon_0/h\nu$, is found in one critical harmonic oscillator. If this has happened, the molecule will have the correct distribution of vibrational energy for reaction to occur.

4.10.3 The Kassel quantum calculation

Step 1. *This requires calculation of the fraction of molecules which have n quanta corresponding to an energy $\varepsilon' = nh\nu$*

$$\left(\frac{k_1}{k_{-1}}\right)_{nh\nu} = \left\{1 - \exp\left(-\frac{h\nu}{kT}\right)\right\}^s \frac{(n+s-1)!}{n!(s-1)!} \exp\left(-\frac{nh\nu}{kT}\right) \tag{4.69}$$

which is the quantum analogue of Equation 4.48.

Step 2. *This calculates the probability that a particular oscillator has at least m quanta, where $\varepsilon_0 = mh\nu$ while all of the s oscillators have between them n quanta where $\varepsilon' = nh\nu$*

$$W = \frac{n!(n-m+s-1)!}{(n-m)!(n+s-1)!} \tag{4.70}$$

which is the quantum analogue of Equation 4.49.

Step 3. *This formulates the steady state treatment of the classical form (Section 4.9.4) in quantum form*

$$k_{2_{(\varepsilon'=nh\nu)}} = k_3 \frac{n!(n-m+s-1)!}{(n-m)!(n+s-1)!}. \tag{4.71}$$

This is the quantum analogue of Equation 4.56.

Step 4. *This calculates the quantum versions of k_{obs}^{1st} and k_∞*

In the classical version the total rate was found by summing the rates for each particular energy ε' in the range $\varepsilon' \to \varepsilon' + \delta\varepsilon$, and the sum was then replaced by an integral over all energies ε_0 to ∞. In the quantum version, the final summation is over all quantum states n, where $n = \varepsilon'/h\nu$ in the range m quanta to ∞, where $m = \varepsilon_0/h\nu$.

$$k_{obs}^{1st} = \sum_{n=m(=\varepsilon_0/h\nu)}^{\infty} \frac{\left\{1 - \exp\left(-\dfrac{h\nu}{kT}\right)\right\}^s \dfrac{(n+s-1)!}{n!(s-1)!} \exp\left(-\dfrac{nh\nu}{kT}\right) k_3 \left\{\dfrac{n!(n-m+s-1)!}{(n-m)!(n+s-1)!}\right\}}{1 + \dfrac{k_3}{k_{-1}[A]} \left\{\dfrac{n!(n-m+s-1)!}{(n-m)!(n+s-1)!}\right\}} \tag{4.72}$$

This is the quantum analogue of Equations 4.63 and 4.64.

$$k_\infty = \sum_{n=m(=\varepsilon_0/h\nu)}^{\infty} \left\{ 1 - \exp\left(-\frac{h\nu}{kT} \right) \right\}^s \frac{(n+s-1)!}{n!(s-1)!} \exp\left(-\frac{nh\nu}{kT} \right) k_3 \left\{ \frac{n!(n-m+s-1)!}{(n-m)!(n+s-1)!} \right\}$$

(4.73)

which is the quantum analogue of Equations 4.65, 4.66.
The summation gives

$$k_\infty = k_3 \exp\left(-\frac{mh\nu}{kT} \right).$$

(4.74)

Since $\varepsilon_0 = mh\nu$ this can be written as

$$k_\infty = k_3 \exp\left(-\frac{\varepsilon_0}{kT} \right)$$

(4.75)

which is identical with the classical value, Equation 4.68.

4.10.4 Comparison of the quantum theory with experiment

1. The values of m and ν are chosen so that $\varepsilon_0 = E_A$, the experimental activation energy, where $\varepsilon_0 = mh\nu$. Having chosen one of m and ν, the other is fixed by the observed value of E_A. The value of ν is taken to be the mean value of all the vibrational frequencies.
2. s is taken to be the total number of vibrational modes.
3. There is no explicit value of k_3, and it is taken to be equal to the experimental A factor found from the dependence of k_∞ on temperature.
4. However, despite points 1, 2 and 3 the quantum version gives a very good fit with experiment, and is a *vast improvement* on the classical version.
5. Like the classical theory it predicts that the plot

$$\frac{1}{k_{obs}^{1st}} \quad \text{against} \quad \frac{1}{[A]}$$

should be a curve, and that the plot

$$\log k_{obs}^{1st} \quad \text{against} \quad \frac{1}{T}$$

should be linear. Both these predictions are verified by experiment.

4.10.5 Points in Kassel's theory which need to be reassessed

1. There is a fundamental inconsistency in the assumption that vibrations are harmonic oscillators:
(a) the *calculation requires* harmonic oscillators which means that there can be no flow of energy between the oscillators.
(b) the *mechanism requires* sufficient anharmonicity so that a flow of energy into the critical oscillator can take place.

2. The same frequency is used for all the oscillators. This is physically unrealistic especially as the activated molecule A*, and the special activated molecule $A^{\otimes}$ must have different distributions of energy.

3. A complete theory would calculate ε_0 and k_3 explicitly. k_3 is the rate constant for passage through the critical configuration of the activated complex, and is as fundamental a quantity as $k_1(\varepsilon')$.

5

Advanced Unimolecular Theory

5.1 Marcus' theory, or Rice, Ramsperger, Kassel, Marcus (RRKM) theory

This theory is a major extension of Kassel's theory. It is a quantum theory which is physically more realistic than Kassel's theory, and it considers the vibrational behaviour of reactant A, of activated molecule A* and of the activated complex A$^{\neq}$ in detail. It deals explicitly with the rate of passage through the critical configuration, and is very interested in what happens physically as an activated molecule A* passes into the activated complex A$^{\neq}$. It also incorporates a quantum version of transition state theory.

5.1.1 New concepts and terminology used in RRKM theory

Two processes are referred to explicitly in the theory

(a) Firstly, there is transfer of energy by collision from translation in one molecule to specific degrees of freedom in the molecule being activated to A*. The theory distinguishes between two types of energy transfer,

   (i) transfer to some degrees of freedom does not contribute to activation,

   (ii) transfer to other degrees of freedom can count towards activation energy.

(b) Once A* is formed, energy flows around the molecule until A* has the correct distribution of energy to be able to pass into the critical configuration. The energy flow has restrictions placed on it - only some degrees of freedom can redistribute their energy, and only some can accept such redistributed energy.

(c) The energy involved in processes (a) and (b) must be precisely defined.

Fixed energy

This is energy which does *not* contribute to accumulation of activation energy, and is *not* able to be redistributed around the activated molecule A*. The degrees of freedom associated with *this type* of energy are *adiabatic* (where the term adiabatic is used, not in the thermodynamic sense, but in the quantum sense of meaning that the quantum states remain unchanged in an adiabatic process). Fixed energy can be:

1. Zero-point energy must always be present, and at the same value, and so can neither be accumulated nor redistributed,

2. Overall translational movement in space cannot be transferred to other modes *within* the same molecule,

3. If a rotational degree of freedom stays in the *same* quantum state, as a consequence of the conservation of angular momentum during an exchange of energy or during redistribution of energy, then energy in this mode cannot be counted towards accumulation of activation energy, and cannot be redistributed around the molecule.

Non-fixed energy

This is the energy which does count to the accumulation of activation energy, and which can be redistributed around the molecule in the process taking the activated molecule A* into the activated complex A*. Non-fixed energy can be:

1. Vibrational energy (other than zero point energy),

2. Rotational energy (other than that which stays in the same quantum state during the particular process),

3. Internal rotations,

4. The free internal translation along the reaction coordinate at the critical configuration. This corresponds to passage through the critical configuration at the moment of chemical transformation.

The degrees of freedom corresponding to non-fixed energy are called "active".

The following diagram refers to non-fixed energy only, Figure 5.1, and describes the energy relations between an activated molecule A* and the activated complex A*. A* is an activated molecule with total non-fixed energy ε', and the ground state of A is taken to be the zero of energy.

The following quantities are specified in Figure 5.1.

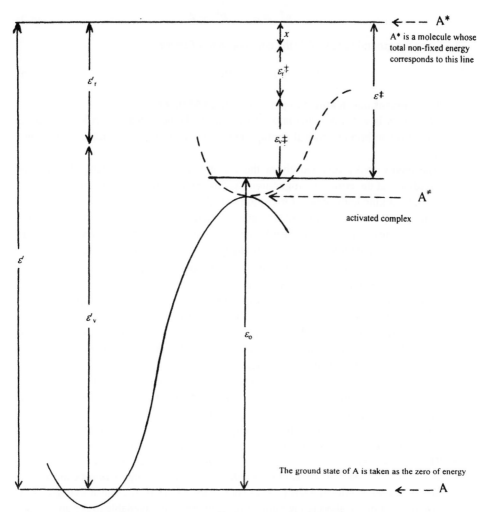

Figure 5.1 Relations between various non-fixed energies for molecule A, where A has total non-fixed energy ε'. The ground state of A is the zero of energy.

1. ε_0 is the difference in energy between the ground state energies of $A^{\neq}$ and A (closely related to the experimental activation energy).

2. ε' is the *total non-fixed energy* of A* where $\varepsilon' \geq \varepsilon_0$. ε' is made up of vibrational and rotational *non-fixed energy*. This energy is in active modes of A*, and has contributed to activation energy, and can be redistributed around A*.

$$\varepsilon' = \varepsilon'_v + \varepsilon'_r. \tag{5.1}$$

3. $\varepsilon^{\neq}$ is the *total non-fixed energy* which the activated complex A has in excess of ε_0. This energy is made up of vibrational and rotational *non-fixed* energy, and includes non-fixed energy x which corresponds to the *internal translation* along the reaction coordinate at the critical configuration

$$\varepsilon^{\neq} = \varepsilon_v^{\neq} + \varepsilon_r^{\neq} + x. \tag{5.2}$$

4. Since the ground state of A is the common zero of energy

$$\varepsilon' = \varepsilon_0 + \varepsilon^{\neq} \qquad \text{or} \quad \varepsilon_0 = \varepsilon' - \varepsilon^{\neq}. \tag{5.3}$$

5.1.2 Important new features appearing in RRKM theory

1. The theory is very interested in what happens as the activated molecule A* changes configuration into the critical configuration corresponding to the activated complex $A^{\neq}$.
2. The theory looks very closely at the internal free translation along the reaction coordinate at the critical configuration. At this stage of the theory it is useful to have a good working knowledge of transition state theory.
3. The theory examines in detail the redistribution of energy which converts A* into $A^{\neq}$. The different possible distributions of energy between non-fixed rotations, vibrations and the internal translation along the reaction coordinate are an important factor in determining the number of quantum states corresponding to ε' and $\varepsilon^{\neq}$, and hence to the equilibrium concentrations of A* and $A^{\neq}$.
4. Non-fixed energy of active vibrations and rotations is subject to rapid statistical redistribution which occurs in times of 10^{-11}s or less.
5. Collisions are assumed to be "strong". This means that large amounts of energy, at least 20 kJ mol^{-1}, are transferred in molecular collisions, and hence activation and deactivation are regarded as *single step* processes.
6. For the high pressure region the theory assumes equilibrium between activated complexes, $A^{\neq}$, and reactants, A. This is a transition state theory assumption, likely to be justified when $\varepsilon_0 \gg kT$, which is generally the case.
7. The theory rests very heavily on being able to calculate the number of ways of distributing energy among the various degrees of freedom of the molecule.
8. The theory has been modified to allow *inactive* modes to have their energy redistributed at random among the *inactive* modes. This energy *cannot* contribute to accumulation of critical energy, or to conversion of A* to $A^{\neq}$. But it *can affect* the rate of reaction by altering the number of quantum states available to A* and $A^{\neq}$ at a given energy.

5.1.3 The RRKM mechanism

This mechanism explicitly considers two important features of reaction. Firstly, the process of changing an activated molecule A* into the activated complex $A^{\neq}$ is described, and this is followed by explicit reference to the rate of passage through the critical configuration as $A^{\neq}$ is converted to products

$$
\begin{array}{lll}
 & k_1(\varepsilon') & \\
\text{A} + \text{A} & \rightarrow & \text{A}^* + \text{A} \\
 & k_{-1} & \\
\text{A}^* + \text{A} & \rightarrow & \text{A} + \text{A} \\
 & k_2(\varepsilon') & \\
\text{A}^* & \rightarrow & \text{A}^{\neq} \\
 & k^{\neq} & \\
\text{A}^* & \rightarrow & \text{products}
\end{array}
$$

There is one sequence of these four steps for each value of the energy ε', so that a whole series of such steps are required when all energies ε' from ε_0 to ∞ are considered.

The rate constants are defined in the following ways.

1. $k_1(\varepsilon')$ is a function of energy, and is explicitly a quantum statistical mechanical quantity.

2. k_{-1} is single valued and independent of energy, and is equal to λZ as in previous theories.

3. $k_2(\varepsilon')$ is the rate constant for formation of the activated complex. It is a quantum statistical mechanical quantity and is a function of energy.

4. k^* is the rate constant for passage through the critical configuration, and is calculated on a transition state theory basis.

5.1.4 The RRKM model
Again this is best described in terms of criteria for activation and reaction. However, in this theory a statement as to the nature of the activated complex is also given.

5.1.5 Criterion of activation
An activated molecule A* has a non-fixed energy ε' in its active modes, where $\varepsilon' \geq \varepsilon_0$.

5.1.6 Criterion of reaction
This energy ε' is redistributed around the active modes of the molecule, until the correct distribution of energy in the correct quantum states is attained, and the critical configuration and energy distribution of the activated complex is formed.

It is unlikely that the initial energy distribution and quantum states of A* immediately after activation will be precisely the energy distribution and quantum states of the critical configuration. Even if this relatively rare quantum state were initially formed, reaction would not be instantaneous since the vibrational modes would have to be correctly phased. A* molecules have lifetimes much greater than the period of vibration (10^{-13}s). They are in the range 10^{-6} to 10^{-11}s.

5.1.7 The activated complex
In earlier theories this did not have to be defined. However, in RRKM theory the activated complex is explicitly considered. The activated complex sits at the minimum maximum potential energy on the potential energy surface, and has all the properties of the activated complex of transition state theory. In particular, movement through the critical configuration corresponds to an internal free translation along the flat portion of the reaction coordinate. Marcus' activated complex is, however, a more refined concept and its energy distributions and quantum states have to be very precisely defined.

5.1.8 The RRKM calculation
Again this calculation can be described in distinct stages.

Step 1. The values of $k_1(\varepsilon')/k_{-1}$ for the range $\varepsilon' \to \varepsilon' + \delta\varepsilon$ are calculated.
The activation/deactivation process is

$$A + A \underset{k_{-1}}{\overset{k_1(\varepsilon')}{\rightleftharpoons}} A* + A$$

and the equilibrium ratio ([A*]/[A] for activation to a given short range of energies $\delta\varepsilon$ must be found, for the case where ε' is non-fixed energy and has the possible values $\infty >$ $\varepsilon' \ge \varepsilon_0$.

Equilibrium statistical mechanics is again used, but this time the expressions are all given in terms of partition functions, f, which are expressible in terms of an exponential energy term and degeneracies g_i and where g_i is the number of quantum states for the given ε'.

$$f = \sum_i g_i \exp\left(-\frac{\varepsilon_i}{kT}\right) \tag{5.4}$$

where the sum is over all energies ε_i.

This is in contrast to transition state theory where the molecular partition function, $Q = f/V$ is used, and where explicit algebraic expressions for f of the type given in Table 2.1 are used. Each of the expressions for f is equal to the corresponding expression (5.4) which involves degeneracies and an exponential energy term.

A common zero of energy is used for A, A* and A^*, and this sometimes results in cancellation of the exponential terms. The common zero is the ground state of A, Figure 5.1.

These considerations give

$$\frac{k_1(\varepsilon' \to \varepsilon' + \delta\varepsilon)}{k_{-1}} = \frac{Q(A*_{\varepsilon' \to \varepsilon' + \delta\varepsilon})}{Q(A)} \tag{5.5}$$

where both of the Q's are that part of the partition function which corresponds to **non-fixed** energy. Each Q is a molecular partition function per unit volume, f/V. This is discussed more fully in Chapter 2, Sections 2.4.4 - 2.4.10.

There is no exponential term here since a common zero of energy is used, giving cancellation

$$\frac{k_1(\varepsilon' \to \varepsilon' + \delta\varepsilon)}{k_{-1}} = \frac{\displaystyle\sum_{\varepsilon' \to \varepsilon' + \delta\varepsilon} g_i' \exp\left(-\frac{\varepsilon'}{kT}\right)}{Q(A)} \tag{5.6}$$

where $Q (A*\varepsilon' \to \varepsilon' + \delta\varepsilon)$ is given in terms of degeneracies and an exponential energy term rather than the explicit algebraic expressions given in Table 2.1
and where
(1) g_i' is the number of possible quantum states for A* in the range
      $\varepsilon' \to \varepsilon' + \delta\varepsilon$
(2) $Q(A)$ is the total partition function per unit volume for reactant A.

Step 2.   This step involves a calculation of $k_2(\varepsilon')$.
          This also can be split up into several steps.
(a)   The rate constant for formation of the activated complex from the activated molecule must be found.
(b)   Firstly, an expression for $k_2(\varepsilon')$ and k^* is obtained from a steady state treatment on the RKKM mechanism.
(c)   This is then converted to an expression involving the equilibrium ratio $[A^*]/[A*]$.
(d)   The non-fixed energy of A^* is made up of contributions from vibrational and rotational modes, and from the internal free translation along the reaction coordinate at the critical configuration. All possible ways of making up this total non-fixed energy of A from internal translational energy and vibrational-rotational energy must be considered.

The steady state treatment

$$\frac{d[A*]}{dt} = k_1(\varepsilon')[A]^2 - k_{-1}[A*][A] - k_2(\varepsilon')[A*] = 0$$

$$[A*] = \frac{k_1(\varepsilon')[A]^2}{k_{-1}[A] + k_2(\varepsilon')} \tag{5.7}$$

$$\frac{d[A^*]}{dt} = k_2(\varepsilon')[A*] - k^*[A^*] = 0$$

$$k_2(\varepsilon') = k^* \frac{[A^*]}{[A*]}. \tag{5.8}$$

These general equations are made more explicit by referring to the range of energy involved, $\varepsilon' \rightarrow \varepsilon' + \delta\varepsilon$

$$k_2(\varepsilon')_{\varepsilon' \rightarrow \varepsilon' + \delta\varepsilon} = \left\{ k^* \left(\frac{[A^*]}{[A*]} \right)_{\text{steady state}} \right\}_{\varepsilon' \rightarrow \varepsilon' + \delta\varepsilon}$$

$$= \left\{ \frac{k^*}{2} \left(\frac{[A^*]}{[A*]} \right)_{\text{equilibrium state}} \right\}_{\varepsilon' \rightarrow \varepsilon' + \delta\varepsilon} \tag{5.9}$$

The factor 2 comes in because equilibrium is assumed between

$$A* \rightleftharpoons A^*$$

and because only those A^* which are moving from left to right through the critical configuration are included.

The expression given above for $k_2(\varepsilon')$ holds for energy in the given short range $\varepsilon' \to \varepsilon' + \delta\varepsilon$. But the Marcus theory considers all possible distributions of energy ε' in the range $\varepsilon' \to \varepsilon' + \delta\varepsilon$.

It is at this stage that the essential features of Marcus' theory start to show themselves. Marcus is very interested in the precise physical properties of the activated molecule and the activated complex, and his theory was developed in a manner calculated to find out just how these precise details affect the final rate expressions. The Marcus theory is interested in much more than the total energy which the activated complex possesses, and is particularly concerned with the flow of energy around the internal modes of the activated molecule. This energy is passed around until the distribution of energy is *precisely* that corresponding to the activated complex. This means that the calculation of $k_2(\varepsilon')_{\varepsilon \to \varepsilon + \delta\varepsilon}$ has to be carried out in such a manner as will recognise the need to allow for consideration of the distribution of non-fixed energy ε' in the activated molecule A^*, and of the distribution of the non-fixed energy $\varepsilon^{\ddagger}$ of the activated complex $A^{\ddagger}$ between vibrational-rotational energy and the energy associated with the internal free translation along the reaction coordinate. It also means considering the quantum states of the activated molecule and the activated complex.

$k_2(\varepsilon')_{\varepsilon \to \varepsilon + \delta\varepsilon}$ therefore covers **all possible distributions** of energy from one limiting situation, $\varepsilon^{\ddagger}_{vr} = 0$, when all of the energy of the activated complex is in the internal free translation, to the other limiting situation when all of the energy of the activated complex is vibrational-rotational, with a zero contribution from the internal free translation. This latter situation makes a zero contribution to $k_2(\varepsilon')$, and is simply the upper limit. A very small, but non-zero energy from the internal translation will correspond to the maximum possible value of $\varepsilon^{\ddagger}_{vr}$ below $\varepsilon^{\ddagger}$. If all possible distributions of energy ε' in the range $\varepsilon' \to \varepsilon' + \delta\varepsilon$ are now considered, this means summing over all distributions from $\varepsilon^{\ddagger}_{vr} = 0$ to $\varepsilon_{vr} = \varepsilon^{\ddagger}$.

$$k_2(\varepsilon')_{\varepsilon' \to \varepsilon' + \delta\varepsilon} = \sum_{\varepsilon^{\ddagger}_{vr} = 0}^{\varepsilon^{\ddagger}} \frac{k^{\ddagger}(x)}{2} \left(\frac{[A^{\ddagger}]_{\varepsilon^{\ddagger}_{vrx}}}{[A^*]} \right)_{equilibrium} \qquad (5.10)$$

where

(i)   the sum holds for energies in the range $\varepsilon' \to \varepsilon' + \delta\varepsilon$

(ii) $\varepsilon' = \varepsilon^{\ddagger} + \varepsilon_0$,

(iii) $\varepsilon^{\ddagger} = \varepsilon^{\ddagger}_{vr} + x$, where x is the energy associated with the internal translation and equals $\frac{1}{2} \mu^{\ddagger} v^2$ in transition state theory,

(iv) $k^{\ddagger}$ depends only on x, but $[A^{\ddagger}]_{\varepsilon^{\ddagger}_{vrx}}$ depends on $\varepsilon^{\ddagger}_{vr}$, x and $\delta\varepsilon$

(v)   each value of $[A^{\ddagger}]_{\varepsilon^{\ddagger}_{vrx}}$ in the summation is an equilibrium concentration for a given value of vibrational-rotational energy, and thus for a given value of x in a range of extent $\delta\varepsilon$. There is thus a whole range of all possible combinations of vibrational-rotational energy, and energy associated with the internal free translation.

Step 3.   *This step calculates k^* which appeared in step 2.*

k^* is a rate constant describing passage through the critical configuration. The derivation is exactly as in transition state theory, Chapter 2, which should now be referred to as only the result of the calculation is quoted

$$k^*(x) = \left(\frac{2x}{\mu^{\neq}}\right)^{\frac{1}{2}} \frac{1}{\delta} = \left(\frac{2x}{\mu^{\neq}\delta^2}\right)^{\frac{1}{2}} \tag{5.11}$$

$$\text{where } x = \frac{1}{2}\mu^* v^2 \text{ and } k^*(x) = v/\delta$$

where $\mu^{\neq}$ is the reduced mass for motion along the reaction coordinate,
and δ is the length along the reaction coordinate defining the critical configuration.

Step 4.   *This completes the calculation of $k_2(\varepsilon')$ and involves finding $[A^{\neq}]_{\varepsilon vr x}/[A^*]_{\varepsilon'}$*

This is found in terms of partition functions which can be expressed in terms of exponential energy terms and degeneracies, where these degeneracies correspond to numbers of quantum states. The arguments leading up to the final expression are complex, algebraic forms for the quantities appearing are specially chosen to allow for cancellation of terms, and the final summation of degeneracies for A does not include all the terms which normally would appear in a straightforward equilibrium expression.

The theory assumes equilibrium between

$$A^* \rightleftharpoons A^{\neq}$$

which means that equilibrium quantum statistical mechanics can be used to calculate the required equilibrium quotient $[A^{\neq}]/[A^*]$ in the short range.

$$\frac{\left\{\begin{array}{l}\left[A^{\neq}\right] \text{ such that } \varepsilon_{vr}^{\neq} \text{ takes} \\ \text{a particular value and } x \text{ lies} \\ \text{in the short range } x \rightarrow x + \delta\varepsilon\end{array}\right\}}{\left\{\begin{array}{l}\left[A^*\right] \text{ such that } \varepsilon' \text{ lies in} \\ \text{the short range } \varepsilon' \rightarrow \varepsilon' + \delta\varepsilon\end{array}\right\}} = \frac{\left\{\begin{array}{l}Q^{\neq} \text{ such that } \varepsilon_{vr}^{\neq} \text{ takes a} \\ \text{particular value and } x \text{ lies} \\ \text{in the short range } x \rightarrow x + \delta\varepsilon\end{array}\right\}}{\left\{Q^*\varepsilon' \rightarrow \varepsilon' + \delta\varepsilon\right\}} \tag{5.12}$$

$$\frac{\left\{\begin{array}{l}Q^{\neq} \text{ such that } \varepsilon_{vr}^{\neq} \text{ takes} \\ \text{a particular value and } x \text{ lies} \\ \text{in the short range } x \to x + \delta\varepsilon\end{array}\right\}}{\left\{\begin{array}{l}Q_{\varepsilon' \to \varepsilon' + \delta\varepsilon}^{*}\end{array}\right\}} = \frac{\left\{\begin{array}{l}\text{a sum of degeneracies calculated in a special way for all} \\ A^{\neq} \text{ for which } \varepsilon_{vr}^{\neq} \text{ takes a particular value and } x \text{ lies in the} \\ \text{short range } x \to x + \delta\varepsilon\end{array}\right\}}{\left\{\begin{array}{l}\text{a sum of degeneracies for } A^* \text{ lying in} \\ \text{the given short range}\end{array}\right\}}$$

The exponential energy terms, exp $(-\varepsilon'/kT)$ for each partition function $Q^{\neq}$ and Q^* have cancelled out because they relate to the same energy.

The denominator:
This is a sum of degeneracies for A^* in the short range $\delta\varepsilon$ and can be written as:

$$\sum_{\text{all } i \text{ in the range } \varepsilon' \to \varepsilon' + \delta\varepsilon} g_i'$$

and this summation is a summation over numbers of quantum states for A^* in the range $\varepsilon' \to \varepsilon' + \delta\varepsilon$

$$\sum_{\text{all } i \text{ in the range } \varepsilon' \to \varepsilon' + \delta\varepsilon} g_i' = \begin{array}{l}\text{total number of quantum states for} \\ A^* \text{ in the range } \varepsilon' \to \varepsilon' + \delta\varepsilon\end{array} \qquad (5.13)$$

To carry out this calculation requires a complete vibrational analysis for the activated molecule A^*.

The numerator
This is much more difficult to formulate, and to put in a precise verbal way just what the sum of degeneracies corresponds to physically.

One of the essential features of Marcus' theory is an explicit consideration of the distribution of non-fixed energy between vibration-rotation and the internal free translation. Each quantum state for each $\varepsilon^{\neq}$ in a range of extent $\delta\varepsilon$ corresponds to a particular value of $\varepsilon_{vr}^{\neq}$, from which the value for x becomes automatically determined since

$$\varepsilon^{\neq} = \varepsilon_{vr}^{\neq} + x. \qquad (5.14)$$

If $\varepsilon_{vr}^{\neq}$ is an explicit value then the resulting value of x must lie in the short range $\delta\varepsilon$.
The numerator thus becomes:

$$\left\{ \begin{array}{l} \text{a sum of degeneracies over} \\ \text{all i for which } \varepsilon_{vr}^{\neq} \text{ takes the} \\ \text{particular value in the range} \\ \text{but } \textit{degeneracies so far} \\ \textit{as x is concerned are not} \\ \textit{included} \end{array} \right\} \quad \text{x} \quad \left\{ \begin{array}{l} \text{a sum of degeneracies for } x \text{ where } x \\ \text{takes a value } \varepsilon^{\neq} - \varepsilon_{vr}^{\neq} \text{ in a range of} \\ \text{extent } \delta\varepsilon, \text{ } \textit{this being the term} \\ \textit{missed out in term A} \end{array} \right\}$$

                Term (A)   Term (B)

Expressions for (A) and (B) must now be given

1. **(A) can be written as**

$$\sum_{\substack{i \text{ for the particular} \\ \text{value of} \varepsilon_{vr}^{\neq}}} g_i^{\neq *}$$

with the physical meaning of this sum as given in the brackets (A), and where the symbol * means that the quantity $g_i^{\neq *}$ is *not* the total degeneracy which would appear in the total partition function for the activated complex, but is a quantity with a *term struck out*. This term relates to the internal translation, and appears in the brackets (B). And so

$$(A) \equiv \sum_{\substack{i \text{ for the particular} \\ \text{value of } \varepsilon_{vr}^{\neq}}} g_i^{\neq *} = \quad \begin{array}{l} \text{the total number of vibrational-rotational} \\ \text{states for which} \varepsilon_{vr}^{\neq} \text{takes this particular value} \end{array}$$

2. **(B) is the number of x states** for which x (the energy associated with the internal translational energy) is in the range $\varepsilon' \rightarrow \varepsilon' + \delta\varepsilon$. The calculation of this quantity is based on a fundamental axiomatic statement of statistical mechanics which allows translation between the formalism of classical mechanics, energy continuously variable, and quantum mechanics, energy discrete. This states that:

> "when there are states so closely spaced in energy that their relevant quantities can be thought of as continuously variable, then the number of states for which a momentum, p, and a corresponding coordinate of position, q, lie in given short ranges is equal to $\dfrac{\delta p \delta q}{h}$ "

In this context:

$$p = \mu^{\neq} v \tag{5.15}$$

$$\varepsilon = \frac{1}{2}\mu^{\neq} v^2 = \frac{p^2}{2\mu^{\neq}} \tag{5.16}$$

so that

$$\frac{d\varepsilon}{dp} = \frac{p}{\mu^{\neq}} \tag{5.17}$$

and

$$d\varepsilon = \frac{p}{\mu^{\neq}} dp \tag{5.18}$$

$$dp = \frac{\mu^{\neq}}{p} d\varepsilon . \tag{5.19}$$

For short ranges

$$\delta\varepsilon = \frac{p}{\mu^{\neq}} \delta p \tag{5.20}$$

$$\delta p = \frac{\mu^{\neq}}{p} \delta\varepsilon. \tag{5.21}$$

Hence

the number of states for the given short range of energy $= \dfrac{2\delta p \delta q}{h}$ $\qquad$ (5.22)

where the factor 2 arises because $+p$ and $-p$ correspond to the same value of ε

$\therefore$ the number of states for the given short range of energy $= \dfrac{2\mu^{\neq} \delta q \delta\varepsilon}{hp}$ $\qquad$ (5.23)

p is the momentum corresponding to the internal translation with internal translational energy, x, and since

$$p = (2\mu^{\neq} \varepsilon)^{\frac{1}{2}} \tag{5.24}$$

then

$$p = (2\mu^{\neq} x)^{\frac{1}{2}} \tag{5.25}$$

q is a coordinate of position and so δq corresponds to δ of transition state theory.

$$\begin{matrix} \text{the number of } x \text{ states} \\ \text{for which } x \text{ is in the} \\ \text{short range of energy, } \delta\varepsilon \end{matrix} = \frac{2\mu^{\neq} \delta}{h(2\mu^{\neq} x)^{\frac{1}{2}}} \delta\varepsilon \tag{5.26}$$

$$= \left(\frac{2\mu^{\neq} \delta^2}{h^2 x} \right)^{\frac{1}{2}} \delta\varepsilon. \tag{5.27}$$

Term B thus $\equiv \left(\dfrac{2\mu^{\neq} \delta^2}{h^2 x} \right)^{\frac{1}{2}} \delta\varepsilon.$

Equation 5.10 can now be written as

$$k_2(\varepsilon')_{\varepsilon' \to \varepsilon' + \delta\varepsilon} = \sum_{\varepsilon_{vr}^{\neq}=0}^{\varepsilon^{\neq}} \frac{k^{\neq}(x)}{2} \left\{ \frac{\displaystyle\sum_{\substack{\text{all } i \text{ for the} \\ \text{particular value } \varepsilon_{vr}^{\neq}}} g_i^{\neq*} \times \left(\frac{2\mu^{\neq}\delta^2}{h^2 x} \right)^{1/2} \delta\varepsilon}{\displaystyle\sum_{\substack{\text{all } i \text{ in the range} \\ \varepsilon' \to \varepsilon' + \delta\varepsilon}} g_i'} \right\}$$

(5.28)

where g_i' relates to A*.

$$\text{But} \quad k^{\neq}(x) = \left(\frac{2x}{\mu^{\neq}\delta^2} \right)^{1/2}$$

(5.29)

$$k_2(\varepsilon')_{\varepsilon' \to \varepsilon' + \delta\varepsilon} = \sum_{\varepsilon_{vr}^{\neq}=0}^{\epsilon^{\neq}} \left\{ \frac{1}{2} \left(\frac{2x}{\mu^{\neq}\delta^2} \right)^{1/2} \left(\frac{2\mu^{\neq}\delta^2}{h^2 x} \right)^{1/2} \left(\frac{\displaystyle\sum_{\substack{\text{all } i \text{ for the particular} \\ \text{value of } \varepsilon_{vr}^{\neq}}} g_i^{\neq*}}{\displaystyle\sum_{\substack{\text{all } i \text{ in the range} \\ \varepsilon' \to \varepsilon' + \delta\varepsilon}} g_i'} \right) \delta\varepsilon \right\}.$$

(5.30)

Cancellation removes both $\mu^{\neq}$ and δ. A similar cancellation happens in the derivation of transition state theory. As in transition state theory, this is fortunate because each of these terms includes δ^2 where δ is the length along the reaction coordinate occupied by the critical configuration. Cancellation means that an explicit value to δ does not need to be given.

The final expression for $k_2(\varepsilon')$ (Equation 5.30) eventually becomes, after cancellation and summation as specified,

$$k_2(\varepsilon')_{\varepsilon' \to \varepsilon' + \delta\varepsilon} = \frac{1}{h} \frac{\displaystyle\sum_{\varepsilon_{vr}^{\neq}=0}^{\varepsilon_{vr}^{\neq}=\varepsilon} g_i^{\neq*}(\text{for A}^{\neq})\delta\varepsilon}{\displaystyle\sum_{\substack{\text{all } i \text{ in range} \\ \varepsilon' \to \varepsilon' + \delta\varepsilon}} g_i'(\text{for A*})}.$$

(5.31)

Step 5   *This is the calculation of k_∞*

$$k_\infty = \sum_{\varepsilon'} \frac{k_1(\varepsilon')k_2(\varepsilon')}{k_{-1}}.$$

(5.32)

Substitution for $k_1(\varepsilon')/k_{-1}$ and for $k_2(\varepsilon')$ followed by cancellation, finally gives

$$k_\infty = \frac{kT}{h} \exp\left(-\frac{\varepsilon_0}{kT} \right) \frac{Q^{\neq*}(\text{for A}^{\neq})}{Q(\text{for A})}$$

(5.33)

where $Q^{\neq*}$ is the partition function for the activated complex for non-fixed energy, comprising vibrational-rotational energy, but with the term for the internal translation missed out. Since this motion in the reaction coordinate has been singled out for special treatment, the activated complex has one degree of vibrational freedom less than a normal molecule. There is a direct parallel here with transition state theory, and the only difference is in the manner in which $Q^{\neq*}$ is calculated.

In RRKM theory $Q^{\neq*}$ is given as

$$Q^{\neq*} = \sum_{\varepsilon_{vr}^\neq=0}^{\varepsilon^\neq} g_i^{\neq*} \exp\left(-\frac{\varepsilon_i}{kT}\right)$$

where $\displaystyle\sum_{\varepsilon_{vr}^\neq=0}^{\varepsilon^\neq} g_i^{\neq*}$ is defined as previously, and is found by direct summing of quantum states for the relevant energy levels. This requires a complete vibrational analysis for the activated complex.

Step 6. *This is the final step, and is a calculation of*

$$\left(k_{obs}^{1st}\right)_{calc} = \sum_{\varepsilon'} \frac{k_1(\varepsilon')k_2(\varepsilon')[A]}{k_{-1}[A]+k_2(\varepsilon')} \tag{5.34}$$

$$= \sum_{\varepsilon'} \frac{k_2(\varepsilon')k_1(\varepsilon')/k_{-1}}{1+k_2(\varepsilon')/k_{-1}[A]} \tag{5.35}$$

Substitution gives a complex expression involving an integral which can be evaluated numerically if the distribution of vibrational-rotational energy levels of reactant, activated molecule and activated complex are known, and this includes the number of vibrational-rotational states for reactant, activated molecule and activated complex. Derivation of these quantities requires a complete vibrational analysis for reactant, activated molecule and also for the activated complex. Obtaining the values for the activated complex is difficult, and depends critically on an assumed structure and configuration for the activated complex. The whole procedure is laborious, even with computer facilities.

5.1.9 Modifications needed

1. It was stated earlier that rotational modes which stayed in the same quantum state when A* changed into $A^\neq$ could not contribute to activation energy, or to the redistribution of energy after activation has occurred. However, the moments of inertia, I^* for A* and $I^\neq$ for $A^\neq$, are often different because the dimensions of A* and $A^\neq$ are different, and hence these modes suffer a change in energy as A* changes to $A^\neq$. This energy is usually released into other degrees of freedom, and this can affect $k_2(\varepsilon)$. An approximate modified treatment corrects $k_2(\varepsilon)$ by multiplying it by $f_1^\neq$ and f_1^* which are partition functions for these adiabatic rotational modes in the activated complex and activated molecule. A more detailed treatment is required.

2. Reaction can proceed via several paths which are kinetically equivalent, and the **number** of such paths gives the statistical factor $l_1^{\neq}$, see chapter 3. If this is included the symmetry number must be omitted from the rotational partition function.

If the modifications in (1) and (2) are developed, a more refined calculation of $k_2(\varepsilon')$ gives

$$k_2\left(\varepsilon'\right)_{\varepsilon'\to\varepsilon'+\delta\varepsilon} = \frac{l^{\neq} f_I^{\neq} \displaystyle\sum_{i\varepsilon_{vr}^{\neq}=0}^{\varepsilon^{\neq}} g_i^{\neq *}(\text{for A}^{\neq})\delta\varepsilon}{hf_I^{*} \displaystyle\sum_{\varepsilon'\to\varepsilon'+\delta\varepsilon} g_i'(\text{for A*})}$$

(5.36)

leading to a more refined formulation of k_∞

$$k_\infty = \frac{kT}{h}\frac{l^{\neq} f_I^{\neq}}{f_I^{*}}\exp\left(-\frac{\varepsilon_0}{kT}\right)\frac{Q^{\neq *}(A^{\neq})}{Q(A)},$$

(5.37)

which in turn leads to a more refined calculation of k_{obs}^{1st}.

5.1.10 Consequences of Marcus' theory

1. There is a vastly improved fit to the fall-off region of k_{obs}^{1st} v [A], though there are still problems associated with the low pressure region which arise from the approximate nature of the treatment of adiabatic rotations. Marcus has produced further modifications, and his final theory and extensions by Quack and Troe are currently regarded as the best ones available. Certainly they are the main ones used nowadays to analyse unimolecular reactions. The nearest rival is Slater's theory which will be discussed in Section 5.3.
2. The theory implies that $1/k_{obs}^{1st}$ v $1/[A]$ is curved.
3. The form for k_∞ allows non-Arrhenius behaviour for the high pressure region. Variation of the pre-exponential term with temperature is possible because of the temperature dependence of the partition functions.
4. For the **high pressure** region, Marcus' theory and transition state theory give identical expressions, except for the inclusion of the transmission coefficient in transition state theory.

5.1.11 Emphasis of the theory

Marcus' theory focuses attention on to the details of activation, and **in particular**, on to the rearrangements and redistributions of energy involved in converting the activated molecule into the activated complex. The theory requires that a detailed formulation of the vibrational states is known for molecule A, and for the activated molecule A* since the number of quantum states for A and A* appear explicitly in the expression for k_{obs}^{1st}. These calculations should be a standard procedure. However, the theory also requires a detailed vibrational analysis for the activated complex since the number of quantum states for A^* must be known to calculate k_∞. This is much more tricky, and requires a model for the activated complex structure.

The theory can be formulated in terms of degeneracies or numbers of quantum states Σ g_i, or in terms of partition functions Q. These are merely different formulations of the same thing, and both ways require the vibrational states to be known and summed.

5.2 Adiabatic channel theory: Quack and Troe
This theory extends the Marcus theory by incorporating ways to allow fully for conservation of angular momentum, and these improved versions are special cases of the adiabatic channel theory of Quack and Troe. The version given here makes considerable use of the concepts of generalised transition state theory, microcanonical transition state theory and Marcus' theory. These concepts and their extensions will be highlighted before discussing adiabatic channel theory.

5.2.1 Previous concepts which are inherent in adiabatic channel theory
The potential energy surface gives the potential energies for all conceivable configurations of the "reaction unit", where the potential energy is as defined in Section 6.1.1 and where the potential energy is calculated for a *system of atoms at rest*. This means that each configuration must be thought of as an instantaneously "static" arrangement of atoms which have no vibrational or rotational motion. Obviously the atoms in these arrangements do have these motions. However, rotational motion only includes kinetic energy and so will not contribute. Vibrational motion has both potential and kinetic energy; this potential energy is included in the calculation because potential energy is a property of position not of motion. The kinetic energy of vibration is, however, a property of the motions of the atoms which occur during a vibration. Since the kinetic energy of the nuclei is not included in the calculation because the atoms are assumed to be at rest for each configuration, then this motion, like the rotational motion, is not of interest to the calculation of the potential energy of transition state theory. Of course, both the movement of a vibration and of a rotation can still be considered, and can be superimposed on the potential energy surface, Chapter 6, but this is an extra added feature not an integral property of the surface. Change of configuration then results from alteration of position under the influence of forces corresponding to the potential energy of the configuration.

Each configuration can represent "reaction units" with different total energies where the total energy is given by a horizontal plane through the surface. For every configuration there is, therefore, a distribution of total energy over all "reaction units" with the given configuration. And for a given "reaction unit" with a given total energy there will be a distribution of vibrational, rotational and internal translational energy corresponding to change of configuration over the surface. As *this* "reaction unit" changes configuration it does so at *constant total energy*, but it will do so giving different distributions of vibrational, rotational and internal translational energy. As the configuration changes the "reaction unit" will consequently have open to it different possible combinations of quantum states all corresponding to the same total energy but different distributions of internal energy. Each configuration, therefore, has open to it a distribution of quantum states, and these will change as configuration changes over the potential energy surface.

The surface could be modified so that for each configuration all the possible quantum states open to that configuration for a given total energy could be listed, and this repeated for all possible total energies.

Basic transition state theory does not enquire explicitly into these quantum states - they only appear implicitly via the partition functions in the final rate expression. In basic transition state theory all possible states are simply a series of interlinking configurations which connect reactant and product configurations in a continuous path. All such paths are possible, but the most likely path is that going via the sequence of lowest lying potential energies changing via the lowest lying potential energy maximum - the minimum energy path or reaction coordinate. *To achieve reaction the critical configuration must be reached, and the rate of change of configuration over this critical configuration defines the rate of reaction*.

In generalised transition state theory changes of configuration *along the minimum energy path* are considered in more detail. These configurations are called generalised transition states or generalised activated complexes, with the term transition state or activated complex being reserved for the critical configuration at the col. The theory calculates the number of quantum states associated with each configuration. It does not need to enquire explicitly into what quantum states are open to a given configuration, though this is implicit in the calculation of the total number open to that configuration. The total number of quantum states is calculated for each configuration along the minimum energy path. Where there are low numbers of quantum states there will be a barrier to change of configuration, and the rate of change of configuration through the given generalised transition state will be low. *The lowest rate will, therefore, be associated with the lowest number of quantum states and this will define the rate of reaction. However, to achieve reaction the "reaction unit" must also pass through the critical configuration*.

Marcus' theory considers the quantum states open to an activated molecule, i.e. a reactant molecule with energy $\varepsilon' \geq \varepsilon_0$. It also considers in detail the quantum states open to the activated complex and in particular, looks at all possible distributions of vibrational, rotational and internal translational energies for all activated complexes, and does so over all energies $\varepsilon' \geq \varepsilon_0$. To do this requires carrying out a complete vibrational analysis for the reactant molecule and for the activated complex. This adds a considerable degree of complexity but, nevertheless, the theory does restrict the calculation of the quantum states to two configurations only, and to movement along the minimum energy path.

However, it is possible to consider a much more detailed and complex series of continuous paths over the surface linking configurations in the same quantum state, i.e. with the same quantum numbers. If one specific quantum state is chosen and all configurations on the potential energy surface which have that particular quantum state within their own distribution of quantum states are specified, then it is possible that there are many continuous pathways over the surface which can link configurations which conform to that specification. All these pathways can be compiled. This is then carried out for all possible quantum states on the surface.

Such pathways have a special name, *adiabatic channels* - adiabatic because they link configurations where the quantum state does not change, and channel is now a new terminology instead of path.

As in conventional transition state theory an "effective potential energy" for each configuration must be found and the change of configuration resulting from forces arising from this effective potential energy found. Along the minimum energy path of conventional transition state theory the vibrational potential energy of the "reaction unit" is zero, but at configurations displaced from the minimum energy path the vibrational potential energy is always greater than zero. This is not immediately obvious, but arises in the following way. Along the reaction coordinate the "reaction unit" undergoes a non-free internal translation, except at the col when it becomes a free internal translation, but movement at right angles to the reaction coordinate corresponds to a vibration, the minimum of which vibration occurs on the reaction coordinate and which, therefore, corresponds to zero vibrational potential energy. This is also the case for all vibrations which occur on the n-dimensional surface of polyatomic "reaction units".

This is best seen by taking the 3-dimensional surface as represented in Figure 2.2 and taking slices down through the surface at right angles to the reaction coordinate. These correspond to dividing planes of generalised transition state theory (Section 3.10 and Figures 3.9-3.11). Each slice will correspond to a Morse-type curve for the vibrational mode in the same way as the sides of the surface at the ends of the entrance and exit valleys (Figure 2.2) show the Morse curve for the symmetric stretch of the reactant and product molecules respectively. As stated earlier the minimum for any vibration occurring on the reaction coordinate corresponds to zero vibrational potential energy, and so any point on the sides of the Morse curve corresponds to non-zero vibrational potential energy; the higher up the sides of the surface, i.e. the greater the displacement from the reaction coordinate, the greater the potential energy. And so configurations lying above the minimum energy path or reaction coordinate will correspond to potential energies which are *higher* than those at the point on the reaction coordinate corresponding to the minimum in the particular vibration. These configurations also possess kinetic energy associated with the relevant vibrational mode, along with kinetic energy of rotation corresponding to the particular rotational quantum state being considered. The *adiabatic channel potential* for a given configuration is defined to be the sum of the potential energy of conventional transition state theory, *plus* the vibrational and rotational energies. It is thus equal to the total energy of the configurations *minus* the internal translational energy along the adiabatic channel. This separation out of the internal translational motion and consequent independent consideration of this motion is analogous to the separation out in transition state theory of the internal translational motion along the reaction coordinate.

The configuration of the "reaction unit" then alters under the action of the forces set up by the adiabatic channel potential, and adiabatic channels can be constructed as described earlier.

These adiabatic channels running over the potential energy surface, each of which links configurations in the same quantum state can be classified as *open* or *closed*. A channel which is *open* is one which allows reaction to occur. Along such channels the "reaction unit" has total energy *greater* than the highest adiabatic channel potential and hence all configurations can be reached. A *closed* channel is one where the adiabatic channel potential *can exceed* the total energy and hence some configurations are inaccessible.

This is analogous to the situation in conventional transition state theory where all configurations on the surface are not necessarily attainable by the "reaction unit". Here only "reaction units" which have total energy equal to or greater than the critical energy can ever reach the critical configuration and react.

5.2.2 Adiabatic channels

The instantaneous state of the "reaction unit", or arrangement of all the atoms involved in the reaction, is described completely by

(i)   a distance along the adiabatic channel which is equal to a distance q along the reaction coordinate

(ii) a set of quantum numbers relating to all other aspects of the set of atoms in question.

As discussed in Section 5.2.1 an adiabatic channel corresponds to a conceivable sequence of changes throughout which all of these quantum numbers remain constant.

The *total energy*, ε_{total}, is then expressed as a sum of

(i)   a term, $\varepsilon_a(q)$, which is what the energy would be for a *given value* of q. This term is called the *adiabatic channel potential* and is an effective potential energy for the particular distance q along the particular adiabatic channel being considered

(ii) and a term for motion corresponding to change of q under the action of this effective potential energy $\varepsilon_a(q)$.

For the reasons given in Section 5.2.4 below this effective potential energy $\varepsilon_a(q)$ will depend on q in a manner which qualitatively resembles the behaviour of $\varepsilon(q)$, the potential energy along the minimum energy path.

5.2.3 Potential energy profiles of conventional transition state theory

The *potential energy $\varepsilon(q)$, along the minimum energy path* is given in Figures 5.2 , 5.3 5.4 and 5.5.

Figure 5.2 shows the standard simple profile for reaction between A and BC where the left-hand side corresponds to the system where A is at large distances from BC and the interaction energy is zero, and so the potential energy approaches an asymptote as the distance along the reaction coordinate approaches zero. Likewise, at the right-hand side of the profile the interaction energy approaches zero since this represents the system where AB is now at large distances from C, and so again the potential energy approaches an asymptote.

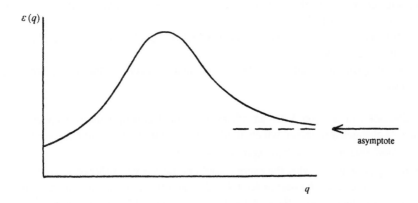

Figure 5.2 A potential energy profile for a bimolecular reaction

Figure 5.3 shows a typical potential energy profile for the unimolecular reaction of isomerisation where one molecule changes into its isomer. The situation is now rather different from that of Figure 5.2 which relates to a bimolecular reaction between two molecules. The left-hand portion of the profile shows the potential energy curve for various configurations of the reactant molecule. Here the left-hand upward part of the curve represents the "squeezing" together of the atoms in the molecule, the first minimum corresponds to $q = q_e$ (reactants), the equilibrium set of internuclear distances for the reactant molecule and the second minimum corresponds to $q = q_e$ (products) found as q increases. q_e's are, therefore, values of q for the reactant and product molecules each in its configuration of lowest potential energy. The potential energies lying between these two minima represent the changes in potential energy as reactant changes into its isomer. Again the extreme right-hand upward part of the curve represents "squeezing" together the atoms in the product isomer.

Figure 5.4 shows a typical decomposition of a reactant molecule into product molecules. Here there is only one minimum, and this corresponds to the reactant molecule. As $q \to \infty$, $\varepsilon(q)$ tends to an asymptote which corresponds to the complete recession of the product molecules from each other, and is very similar to the right-hand part of the profile for a bimolecular reaction, Figure 5.2.

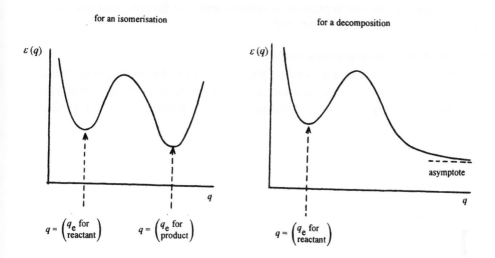

Figure 5.3 caption (for an isomerisation) and Figure 5.4 (for a decomposition):

for an isomerisation

for a decomposition

$\varepsilon(q)$

$\varepsilon(q)$

asymptote

q

q

$q = \begin{pmatrix} q_e \text{ for} \\ \text{reactant} \end{pmatrix}$ $q = \begin{pmatrix} q_e \text{ for} \\ \text{product} \end{pmatrix}$ $q = \begin{pmatrix} q_e \text{ for} \\ \text{reactant} \end{pmatrix}$

**Figure 5.3 A potential energy profile for
an isomerisation**

**Figure 5.4 A potential energy profile for
a decomposition into product
moleucles**

Some decompositions, such as the decomposition of molecules into radicals, for example

$$C_2H_6 \rightarrow CH_3^{\cdot} + CH_3^{\cdot}$$

do not give a maximum in $\varepsilon(q)$ along the minimum energy path but go straight to a plateau, Figure 5.5.

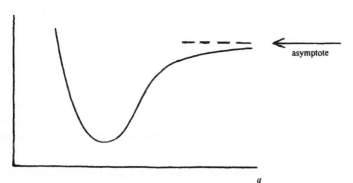

asymptote

q

Figure 5.5 A potential energy profile for a decomposition into radicals

This is a consequence of the zero or near zero activation energy observed for the recombination of most radicals. The two $CH_3^{\cdot}$ radicals end up with approximately the same potential energy as the activated complex on the minimum energy path, proved by finding that the activation energy for decomposition of the molecule is approximately equal to the dissociation energy, D_0. This has the consequence that for a decomposition into radicals it is impossible to place the activated complex on the path, since there is no maximum along the minimum energy path.

5.2.4 Adiabatic channel potential profiles

For a decomposition, the adiabatic channel potential $\varepsilon_a(q)$, can depend on q as illustrated in Figure 5.6 where $\varepsilon_a(q)$ for two different adiabatic channels are compared with $\varepsilon(q)$ for the minimum energy path for a decomposition of a molecule into two product molecules.

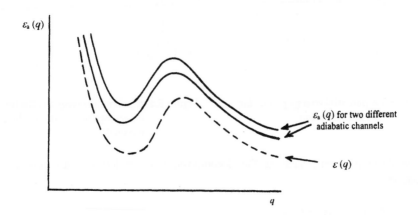

Figure 5.6 Dependence of the adiabatic channel potential on q

Often the general shape is similar, but $\varepsilon_a(q)$ **always** lies above $\varepsilon(q)$ because it includes vibrational and rotational energy corresponding to the specific quantum state involved (Section 5.2.1). However, the maximum in $\varepsilon_a(q)$ is not necessarily at the same value of q as the maximum in $\varepsilon(q)$ because the adiabatic channel normally will not coincide with the minimum energy path, but can be a path over any part of the surface. Furthermore, the adiabatic channel potential depends on the detailed shape of the surface over which the channel occurs. As the configuration changes, the frequency of the vibration can change **even though** the vibrational quantum number does not, and this will result in corresponding variations in the vibrational potential energy. However, $\varepsilon_a(q)$ will not depart drastically from the general shape for $\varepsilon(q)$ unless the vibrational frequencies for the configurations are strongly dependent on q.

For decompositions into radicals, Figure 5.5 the minimum energy path $\varepsilon(q)$ does not show a maximum and the activated complex cannot be placed on the path. The adiabatic channel potential $\varepsilon_a(q)$ can, however, show a maximum, Figure 5.7.

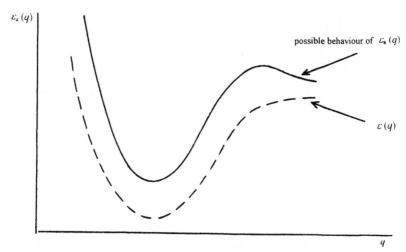

Figure 5.7 A potential energy profile for a decomposition into radicals

This again is a consequence of the fact that an adiabatic channel will link configurations on the surface displaced from the minimum energy path and these may involve configurations where the vibrational frequency and hence the vibrational energy could be high resulting in a maximum for $\varepsilon_a(q)$. If this happens then it will allow the activated complex to be placed along the adiabatic channel in contrast to the impossibility of doing this on the minimum energy path.

5.2.5 Open and closed adiabatic channels
Adiabatic channels are classed as open or closed (Section 5.2.1) in terms of whether a reactant molecule with q initially close to q_e can pass over classically into product without leaving the channel or not. This depends on the total energy, ε_{total}, as well as on the channel, Figures 5.8 and 5.9.

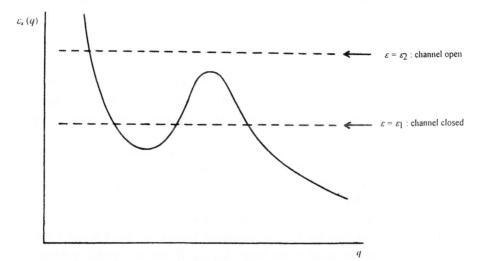

Figure 5.8 Open and closed channels: dependence on total energy

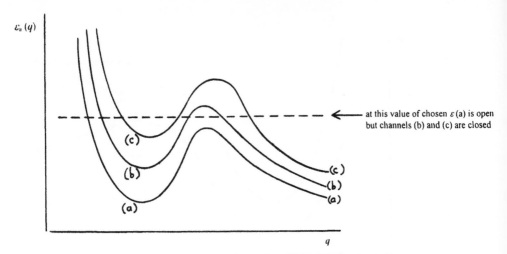

Figure 5.9 Open and closed channels at constant total energy: dependence on channel

In Figure 5.8 when the total energy is greater than the value of the adiabatic channel potential $\varepsilon_a(q)$ at the maximum, the channel will be termed open. If it is less, then the channel is closed.

In Figure 5.9 whether a channel is open or closed will depend on the particular channel. For a given value of the total energy, ε_{total}, as marked, channel a will be open, but channels b and c are closed.

5.2.6 A microcanonical, but non adiabatic, derivation of a general rate constant for passage through a generalised transition state

This derivation relies heavily on the concept of the generalised transition state as well as on the algebraic and statistical mechanical formalism of Marcus. The essence of the Marcus theory is the explicit consideration of the distribution of energy between the vibrational and rotational states and the internal translation along the reaction coordinate at the critical configuration. Quack and Troe's arguments are based on concepts and reasoning used in nuclear physics. Both lines of approach lead essentially to the same conclusions, but the Marcus approach is developed here because it is much more directly linked to the arguments used in previous sections of this chapter and in Section 3.10 of Chapter 3.

The derivation of a general rate constant assumes that only conditions relating to the energy need to be considered. Reactant molecules with energy in the short range $\varepsilon \to \varepsilon + \delta\varepsilon$ enter a particular generalised transition state which covers a short range $q \to q + \delta q$ along the reaction coordinate.

The state of the molecule depends on the internal translational energy for movement along the reaction coordinate as well as on all other contributions to the energy. Attention is focused on the generalised activated complexes which are in the generalised transition state under scrutiny.

The total energy, $\varepsilon = \varepsilon_{vr} + \varepsilon_{int\ trans}$ is in the given short range $\varepsilon \to \varepsilon + \delta\varepsilon$. If one specific quantum state corresponding to ε_{vr} is considered, because ε_{vr} takes a unique value, then the range $\delta\varepsilon$ will be equal to the range $\delta\varepsilon_{int\ trans}$. If this one specific quantum state corresponding to the given values of the quantum numbers v and J is considered, then the number of quantum states open to a generalised activated complex with this set of quantum numbers and with energy in the range $\varepsilon \to \varepsilon + \delta\varepsilon$ will correspond to the number of quantum states resulting from the range $\delta\varepsilon_{int\ trans}$. These will, of course, refer only to the forward moving generalised activated complexes. For these generalised activated complexes the momentum $p = \mu^{\ddagger}v$, for movement over a short range of distances along the reaction coordinate $q \to q + \delta q$ will be restricted to a short range $p + \delta p$, corresponding to the short range in energy $\varepsilon \to \varepsilon + \delta\varepsilon$. The number of quantum states in respect of internal translation will then be $\delta p \delta q / h$.

This is an axiomatic statement and comes from fundamental propositions of statistical mechanics which allow conversion between classical mechanics and quantum mechanics formalism (Section 5.1.8, Step 4). And so,

the number of generalised activated complexes which have the given set of quantum numbers v and J, and have total energy in the range $\varepsilon \to \varepsilon + \delta\varepsilon$	$\propto$	the number of quantum states resulting from the range $\delta\varepsilon_{int\ trans}$

$$= \frac{\delta p\, \delta q}{h}.$$

(5.38)

It then follows, provided there is a local Maxwell-Boltzmann equilibrium, that

$$\frac{\begin{array}{l}\text{the number of these generalised} \\ \text{activated complexes which have} \\ \text{the given set of quantum numbers} \\ v \text{ and } J, \text{ and have total energy} \\ \text{in the range } \varepsilon \to \varepsilon + \delta\varepsilon\end{array}}{\begin{array}{l}\text{the number, } N_{\varepsilon \to \varepsilon + \delta\varepsilon}, \text{ of} \\ \text{reactant molecules which have} \\ \text{total energy in the range} \\ \varepsilon \to \varepsilon + \delta\varepsilon\end{array}}$$

$$=$$

the number of quantum states resulting from the range in the internal translational energy
$$\varepsilon_{int\ trans} \to \varepsilon_{int\ trans} + \delta\varepsilon$$

the number of quantum states of the reactant molecules for which the total energy lies in the range $\varepsilon \to \varepsilon + \delta\varepsilon$

$$= \frac{\delta p\, \delta q}{h}\, \frac{1}{\rho(\varepsilon)\, \delta\varepsilon}.$$

(5.39)

The term $\rho(\varepsilon)\delta\varepsilon$ is the quantity appearing in the right-hand side of Equation 5.13, and is equal to the sum of degeneracies for the generalised activated complexes lying in the range $\delta\varepsilon$, and this in turn is equal to the number of quantum states lying in the short range of energy $\delta\varepsilon$. $\rho(\varepsilon)$, on the other hand, is called the "density of states" of the reactant molecule, and is a standard quantity of statistical mechanics. Physically it corresponds to the number of quantum states per **unit range of energy**, in contrast to $\rho(\varepsilon)\delta\varepsilon$ which is the number of quantum states *lying in a short range of energy*. In the Marcus theory the "sum of degeneracies" in the short range $\delta\varepsilon$ is used throughout without $\rho(\varepsilon)\delta\varepsilon$ or $\rho(\varepsilon)$ necessarily being formally stated. However, in adiabatic channel theory the quantity which appears in the final equations is the "density of states" $\rho(\varepsilon)$ rather than the "sum of degeneracies", $\rho(\varepsilon)\delta\varepsilon$.

And so, by rearrangement of Equation 5.39

The number of these generalised activated complexes which have the given set of quantum numbers and have total energy in the range
$\varepsilon \rightarrow \varepsilon + \delta\varepsilon$

$$= \frac{\delta p\, \delta q}{h} \frac{N_{\varepsilon \rightarrow \varepsilon + \delta\varepsilon}}{\rho(\varepsilon)\delta\varepsilon}.$$

$$(5.40)$$

It is now possible to calculate the rate at which these particular generalised activated complexes will pass on into another particular generalised transition state.

The probability per unit time that a generalised activated complex will pass on into another generalised transition state

$$= \frac{\text{the velocity of passing on}}{\delta q}$$

$$= \frac{p}{\mu\delta q}$$

$$(5.41)$$

where μ is the relevant reduced mass for movement along the reaction coordinate, and $p = \mu v$, giving $v = p/\mu$.

The number of these generalised activated complexes passing on per unit time into another generalised transition state

$$= \left(\frac{p}{\mu\delta q}\right) \times$$

the number of those generalised activated complexes which have the given set of quantum numbers v and J, and have energy in the range ε to $\varepsilon + \delta\varepsilon$

and by substitution of Equation 5.40.

$$= \frac{p}{\mu\delta q} \frac{\delta p\, \delta q N_{\varepsilon \rightarrow \varepsilon + \delta\varepsilon}}{h\rho(\varepsilon)\delta\varepsilon}$$

$$= \frac{p\delta p}{\mu} \frac{N_{\varepsilon \rightarrow \varepsilon + \delta\varepsilon}}{h\rho(\varepsilon)\delta\varepsilon}.$$

$$(5.42)$$

The range of momentum δp is controlled by the condition that the total energy must lie in the range $\varepsilon \rightarrow \varepsilon + \delta\varepsilon$.

$$\text{The total energy } \varepsilon = \varepsilon_{vr} + \varepsilon_{\text{int trans}} \tag{5.43}$$

$$= \varepsilon_{vr} + \frac{1}{2}\frac{p^2}{\mu} \tag{5.44}$$

since $\varepsilon = \frac{1}{2}\mu v^2$ and $v^2 = p^2/\mu^2$. If the total energy ranges from $\varepsilon \rightarrow \varepsilon + \delta\varepsilon$, then $\varepsilon_{vr} + \varepsilon_{\text{int}}$ will range from

$$\varepsilon_{vr} + \frac{p^2}{2\mu} \quad \text{to} \quad \varepsilon_{vr} + \frac{(p+\delta p)^2}{2\mu}$$

and provided that $\delta p \delta q$ is small then $\delta\varepsilon$ corresponds to $p\delta p/\mu$, from which it follows that, by substitution of $p\delta p/\mu$ for $\delta\varepsilon$,

the number of generalised activated complexes which have the given set of quantum numbers v and J, and have total energy in the range $\varepsilon \rightarrow \varepsilon + \delta\varepsilon$,and which pass on per unit time into another generalised transition state.

$$= \frac{N_{\varepsilon\rightarrow\varepsilon+\delta\varepsilon}}{h\rho(\varepsilon)}, \tag{5.45}$$

and this equation refers to all the generalised activated complexes which have the **specified** quantum numbers v and J. Equation 5.45 must, therefore, be multiplied by the number of generalised activated complexes with **all possible** v and J in the range $\varepsilon \rightarrow \varepsilon + \delta\varepsilon$, giving

the number of generalised activated complexes at the particular generalised transition state with total energy in the range $\varepsilon \rightarrow \varepsilon + \delta\varepsilon$ which pass on per unit time

$$= \frac{w(\varepsilon)N_{\varepsilon\rightarrow\varepsilon+\delta\varepsilon}}{h\rho(\varepsilon)} \tag{5.46}$$

where $w(\varepsilon)$ is the number of states in respect of features other than internal translation for which, at the generalised transition state in question, the energy is $\leq \varepsilon_{\text{total}}$. Each of the states $w(\varepsilon)$ makes the same contribution

$$\frac{N_{\varepsilon \to \varepsilon + \delta \varepsilon}}{h\rho(\varepsilon)}$$

to the total number of generalised activated complexes which pass on to the next generalised transition state per unit time.

The total number which
pass on per unit time
per unit volume

$$= \frac{w(\varepsilon)N_{\varepsilon \to \varepsilon + \delta \varepsilon}}{h\rho(\varepsilon)V} \qquad (5.47)$$

$$= \frac{k(\varepsilon)N_{\varepsilon \to \varepsilon + \delta \varepsilon}}{V} \qquad (5.48)$$

$$k(\varepsilon) = \frac{w(\varepsilon)}{h\rho(\varepsilon)} \qquad (5.49)$$

where $k(\varepsilon)$ is a general rate constant for the generalised activated complexes with total energy ε.

5.2.7   A microcanonical derivation of the rate constant for passage through a generalised transition state along an adiabatic channel

The previous section (5.2.6) dealt with the application of generalised transition state theory to a unimolecular reaction where movement along the **minimum energy path** was considered, and this path linked configurations where the quantum numbers can change. If, however, change of configuration is only allowed along an **adiabatic channel** this will severely limit the treatment since only channels linking configurations with the same quantum number will be allowed. This is taken care of in the theory where imposition of the condition is described by condition L.

In this derivation the state of the "reaction unit" depends not just on the distance q along an adiabatic channel, but specifically on the particular set of quantum numbers possessed by the reactant molecules.

If passage through some generalised transition state is considered where the original molecules meet some condition, L, such as having one or more quantum numbers prescribed, then a modification of the above treatment leads to the result

the total number of generalised activated
complexes at the particular generalised
transition state in question, having total
energy in the range $\varepsilon \to \varepsilon + \delta \varepsilon$ and
meeting some condition L, which pass
on to another generalised transition state
per unit time per unit volume

$$= \frac{w_L(\varepsilon)N_{L_{\varepsilon \to \varepsilon + \delta \varepsilon}}}{h\rho_L(\varepsilon)V} \qquad (5.50)$$

$$= \frac{k_L(\varepsilon)N_{L_{\varepsilon \to \varepsilon + \delta \varepsilon}}}{V} \qquad (5.51)$$

$$k_{L}(\varepsilon) = \frac{w_{L}(\varepsilon)}{h\rho_{L}(\varepsilon)}$$

(5.52)

where $N_{L\,\varepsilon\,\to\,\varepsilon+\,\delta\varepsilon}$ is the number of reactant molecules having total energy in the range $\varepsilon \to \varepsilon + \delta\varepsilon$ and meeting condition L.

$w_{L}(\varepsilon)$ is the number of states in respect of features other than internal translation for which, at the generalised transition state in question, the energy other than internal translation energy $\leq \varepsilon_{total}$ and condition L is met.

$\rho_{L}(\varepsilon)\delta\varepsilon$ is the number of states of reactant molecules for which the total energy is in the range $\varepsilon \to \varepsilon + \delta\varepsilon$ and condition L is met and $\rho_{L}(\varepsilon)$ is the number of states per unit range of energy for which condition L is met.

$k_{L}(\varepsilon)$ is an "ultra specific" rate constant for reactant molecules of a given energy ε and which meet condition L.

In these expressions $\rho_{L}(\varepsilon)$ is a property purely of the reactant. It is entirely independent of the part of the surface near the col. $w_{L}(\varepsilon)$ in contrast does depend on this region and, in particular, will vary from one generalised transition state to another.

5.2.8 A microcanonical derivation of the rate of reaction along an adiabatic channel

Section 5.2.7 deduced the rate expression for passage through a given generalised transition state. For reaction to occur it must also pass through *all* subsequent transition states. This can be expressed as

$$w_{L}(\varepsilon) = w_{L}'(\varepsilon) + w_{L}''(\varepsilon)$$

(5.53)

so that

$$w_{L}'(\varepsilon) = w_{L}(\varepsilon) - w_{L}''(\varepsilon)$$

(5.54)

where $w_{L}'(\varepsilon)$ is the number of states for which the conditions listed above in Section 5.2.7 are met, *and all subsequent generalised transition states can be passed through*, i.e. which constitute an open channel and allow reaction to occur.

$w_{L}''(\varepsilon)$ is the number of states for which the conditions in Section 5.2.7 are met, but for which *it is not possible to pass through some of the subsequent generalised transition states*, i.e. which constitute a closed channel and do *not* allow reaction to occur. It is a term limiting $w_{L}'(\varepsilon)$.

If a reactant molecule can react when it is in an open channel but cannot react when it is in a closed channel, then $w_{L}'(\varepsilon)$ and $w_{L}''(\varepsilon)$ can be expressed otherwise as

$w_{L}'(\varepsilon)$ is the number of open channels for reactant molecules for which the total energy is in the range $\varepsilon \to \varepsilon + \delta\varepsilon$ and condition L is met.

$w_{L}''(\varepsilon)$ is the number of closed channels for reactant molecules for which the total energy is in the range $\varepsilon \to \varepsilon + \delta\varepsilon$ and condition L is met.

$w_{L}(\varepsilon)$ is the total number of channels for reactant molecules, open and closed, for which the total energy is in the range $\varepsilon \to \varepsilon + \delta\varepsilon$, and condition L is met.

The fraction

$$\frac{w_L'(\varepsilon)}{w_L} = \frac{w_L'(\varepsilon)}{w_L'(\varepsilon) + w_L''(\varepsilon)} \tag{5.55}$$

is, therefore, the fraction of open channels to the total number of channels considered, both open and closed, and as such, this fraction of generalised activated complexes will pass through the activated complex and end up as products, i.e. react.

If the total number of generalised activated complexes at a particular generalised transition state having total energy in the range $\varepsilon \rightarrow \varepsilon + \delta\varepsilon$ and meeting some condition L which can pass on to another generalised transition state per unit time per unit volume is Equation 5.50

$$\frac{w_L(\varepsilon)N_{L\,\varepsilon \rightarrow \varepsilon + \delta\varepsilon}}{h\rho_L(\varepsilon)V}$$

the total number of reactant molecules having total energy in the range $\varepsilon \rightarrow \varepsilon + \delta\varepsilon$, and meeting condition L, *which react* per unit volume per unit time

$$= \frac{w_L'(\varepsilon)w_L(\varepsilon)N_{L\,\varepsilon \rightarrow \varepsilon + \delta\varepsilon}}{w_L(\varepsilon)h\rho_L(\varepsilon)V}$$

$$= \frac{w_L'(\varepsilon)N_{L\,\varepsilon \rightarrow \varepsilon + \delta\varepsilon}}{h\rho_L(\varepsilon)V}, \tag{5.56}$$

from which

$$k_L(\varepsilon) = \frac{w_L'(\varepsilon)}{h\rho_L(\varepsilon)}. \tag{5.57}$$

What has now to be done is to characterise the adiabatic channels sufficiently so as to enable $w_L'(\varepsilon)$ to be found over the whole range of energies ε which contribute significantly to the rate. The evaluation of $\rho_L(\varepsilon)$ has been largely reduced to a computational procedure, and is seriously affected by anharmonicity of vibrations and non-rigidity of rotations.

5.2.9 Matching of quantum numbers in reactant and product
It is necessary when constructing adiabatic channel potentials to decide which set of quantum numbers for the products "belongs" with a given set of quantum numbers for the reactants. The procedure can be summarised as

(i)   Some vibrational modes in the product (or products) may correspond closely to vibrational modes in the reactant. An adiabatic channel will link an initial state to a final state in which the vibrational quantum numbers for these states remain unchanged.

(ii)   Conservation of angular momentum requires that a similar relationship will apply in relation to some, but not necessarily all, of the quantum numbers concerned with angular momentum.

(iii) Behaviour of other quantum numbers in particular for normal modes in the reactant which have no close analogue in the product, is less straightforward, but there a general procedure is available.

5.2.10 Success of adiabatic channel theory
This theory is a considerable advance on the non-adiabatic Marcus theory. Many of the reactions studied by Quack and Troe demonstrate the necessity for dealing with the ultra specific rate constants $k_L(\varepsilon)$, and application of the theory requires that restrictions due to conservation of angular momentum must be explicitly considered. Many of the problems in the application of adiabatic channel theory lie in the computational aspects.

5.3 Slater's theory
The theories so far discussed are essentially statistical treatments, where the emphasis is placed on the statistical distribution of energy between the normal modes of vibration, and how this affects the rates of the activation and reaction steps.

Slater's theory is entirely different. It looks at what happens to a molecule between collisions, with emphasis on the detailed dynamical behaviour of the decomposing molecule, and it treats in detail the changing positions of the atoms in the molecule as the molecule vibrates, and studies these different arrangements of the atoms as a function of time. This means that all interatomic distances, bond lengths and bond angles are calculated as a function of time for specified energies and energy distributions of the activated molecule. Since molecular vibrations vary periodically, then the bond lengths, bond angles and atomic distances will also change periodically with time. It will be shown later, that the changing atomic positions depend critically on the vibrational energy of the activated molecule, and the way in which this energy is distributed among the normal modes of vibration.

Fixing the magnitude of this energy, and the energy distribution results in an inevitable and totally unique variation of the positions of the atoms with respect to each other, and reaction can only occur if certain atomic dimensions are achieved. What they are depends on the particular chemical reaction, but essentially this boils down to deciding first on the critical configuration, and then focusing attention on to a critical aspect of the critical configuration, such as a bond length, a bond angle or a combination of these. Slater's theory sets out the physical conditions and requirements which have to be met before it can be possible for the activated molecule to reach the critical dimensions.

Originally this was done by treating the normal modes of vibration as classical harmonic oscillators. Slater has also put forward a quantum theory which copes better, but this theory has not been treated as thoroughly as the classical version. Both of these theories treat the normal modes of vibration as *harmonic* oscillators, which means that there can be *no flow of energy* between the normal modes. The distribution of energy between the normal modes therefore *remains fixed between collisions*, and this imposes a very severe restriction on which atomic positions and dimensions are possible for the activated molecule. However, it is well known from spectroscopic work that *energy can flow* around the modes between collisions. Allowing for such an energy flow vastly improves the theory, and gives results more closely resembling the RRKM theory. But again, the original classical harmonic version is what has normally been used. What is really wanted is a quantum non-harmonic version, but the mathematical problems here are prodigious.

Slater's model is physically reasonable and simple in conceptual detail, but this is more than compensated for by the vastly greater mathematical complexity of the theoretical treatment. For this reason only an outline of the manner of treatment can be given here, and the emphasis will be on the physical chemical aspects of the theory.

5.3.1 Slater's Mechanism

This is virtually the same as Lindemann's, but the emphasis is placed on the changes of atomic positions in the molecule after collisions have produced the activated molecule

$$A + A \xrightarrow{\quad k_1(\varepsilon') \quad} A^* + A$$

$$A^* + A \xrightarrow{\quad k_{-1} \quad} A + A$$

$$A^* \xrightarrow{\quad k_2(\varepsilon') \quad} \text{products}$$

Each of these holds for all energies ε' which are greater than ε_0, and the following expressions hold

$$\text{Rate} = \sum_{\varepsilon'} k_2(\varepsilon')[A^*] \tag{4.43}$$

$$= \sum_{\varepsilon'} \frac{k_2(\varepsilon')k_1'(\varepsilon')[A]^2}{k_{-1}[A] + k_2(\varepsilon')} \tag{4.44}$$

$$k_{obs}^{1st} = \sum_{\varepsilon'} \frac{k_2(\varepsilon')k_1'(\varepsilon')/k_{-1}}{1 + k_2(\varepsilon')/k_{-1}[A]} . \tag{4.45 rearranged}$$

$$k_\infty = \sum_{\varepsilon'} \frac{k_1'(\varepsilon')k_2(\varepsilon')}{k_{-1}} . \tag{4.46}$$

5.3.2 Basic assumptions of the classical harmonic Slater theory

1. The version given deals with a polyatomic molecule where the vibrations are classical and strictly harmonic.
2. The theory assumes that the geometrical arrangements of the atoms in the critical configuration can be given, and its dimensions estimated. The decision made here has a profound effect on the result of the Slater calculation.
3. From this estimate of the structure of the activated complex, a critical coordinate is chosen, and reaction is allowed to occur if this critical coordinate exceeds a certain critical magnitude. This critical coordinate could be a bond length, or combination of bond lengths, a bond angle or any linear combination of bond lengths and bond angles, and the decision is influenced by the nature of the chemical reaction

occurring. An example could be the breaking of a bond after it reached a critical extension, as in the decomposition of N_2O

$$N_2O \quad \rightarrow \quad N_2 + O\cdot$$

where an $N = 0$ bond breaks.

3. The theory assumes that all the normal modes of vibration are relevant, though, for some molecules, some modes might not contribute to the particular coordinate chosen to be the critical coordinate. The geometry of the molecule is the critical factor here. For instance in the N_2O decomposition, the bending modes do not contribute to the variation of the $N = 0$ distance which is chosen to be the critical coordinate.

4. Collisions produce the activated molecule, and the total energy of a molecule stays the same between collisions. It is at this stage that Slater places a *very severe physical restriction* on what can happen to the activated molecule. Energy is not allowed to flow around the molecule between collisions, and this means that the activated molecule A* remains in the same vibrational and rotational states until it reacts, or is deactivated by a further collision. Thus the energy distribution of the A* molecule is fixed and constant after activation has occurred. Slater is interested in the details of this energy distribution only in so far as they affect the geometrical arrangements and positions of the atoms with respect to each other. Marcus was interested in the details of the energy distribution because this affected the number of quantum states open to the activated molecule and activated complex.

5. Once activation has happened, various geometrical arrangements of the atoms result from the combined movements of the atoms as described by the various normal modes. The geometrical arrangements vary with time, and are *inevitable*, depending only on the particular energy distribution which the activated molecule attained on activation.

6. Slater looks at which normal modes contribute to the critical coordinate, and uses the results of a vibrational analysis on the molecule to calculate the periodic displacements affecting the atoms in each normal mode. When added together these give the variation in the critical coordinate with time. This requires investigating how the various vibrations come into phase.

7. The time lag depends on the amount of energy which the activated molecule possesses *in excess* of the critical energy ε_0. It is also related to the number of times per unit time that the vibrations contributing to the critical coordinate come suitably into phase to allow the vital critical coordinate defining reaction to reach its critical value, or greater. The Slater theory is primarily concerned with calculating this quantity.

8. The theory assumes that a complete vibrational analysis of the activated molecule A* is known. The theory, however, used data from analysis of an A molecule, that is of a low energy molecule. This assumes that the behaviour of a molecule on the point of reaction can be estimated from the properties of a molecule in a non-excited state.

5.3.3 Slater's Model
Again this is best described by stating what is meant by activation, and what is involved in reaction.

5.3.4 Criterion of activation
A molecule is activated if it has energy at least ε_0 in its total number of normal modes of vibration. Once it is activated, the distribution of energy among the normal modes of vibration remains fixed.

5.3.5 Criterion of reaction
An activated molecule reacts when its normal modes of vibration come suitably into phase to give the vital coordinate defining reaction its critical magnitude or greater. This is more stringent than appears at first sight. *Since energy is not allowed to flow around the molecule*, this requires that when a molecule acquires the critical energy, or greater, it also must have this critical energy distributed among the normal modes in the correct way. Only some distributions of energy will be such that the critical coordinate can reach its critical magnitude. Some molecules may have the required energy in the total number of normal modes, but do not have this energy distributed in a suitable way. These molecules will never be able to decompose, even when the vibrations do come into phase. Other activated molecules will start off with a suitable distribution, and will be able to react because the distribution enables the critical coordinate to attain its critical magnitude. The theory finds out which distributions can lead to reaction, and which cannot. So a molecule A may have enough energy to become an A*, but not have a suitable distribution of energy to become an A*.

5.3.6 Aspects of vibrational analysis relevant to Slater's theory
If vibrations are harmonic, then the overall motions of the atoms in the molecule can be split up into normal modes of vibration, or oscillators. Some normal modes will involve movements of all the atoms in the molecule, while some will be made up of movements from certain specified atoms only. The assignment of which atoms contribute to a given mode is unique. Likewise the sort of motion executed by the normal mode is unique. The overall motions of the atoms in a molecule are a sum over all atoms of all the individual movements of each atom for every normal mode that each atom is involved in. A frequency can be assigned to every normal mode, and each atom involved in the normal mode moves with this frequency.

The displacements of the atoms are reflected in the amplitude of the vibration, and this amplitude depends on the energy of the normal mode, the greater the energy of the normal mode the larger are the amplitudes and the displacements of the atoms contributing to the normal mode. There are different amplitudes and different displacements for each atom in a normal mode, in contrast to the single frequency applicable to the movement of all the atoms in the normal mode.

If the normal mode is harmonic, the periodic motion can be given by a sine or cosine wave. The information required to construct such waves comes from a complete vibrational analysis of the molecule in question.

5.3.7 Relationships between amplitude and energy, and between displacements of atoms and amplitudes

The amplitude, A_i, for a given atom, i, in a given mode is related to the energy of that mode ε.

$$A_i = \alpha_i' \sqrt{\varepsilon} \qquad (5.58)$$

where α_i' is a constant of proportionality, called the amplitude factor, and values of α_i' can be found from a spectroscopic analysis of the molecule.

The displacements, x_i of each atom involved in the normal mode can be calculated once the amplitudes are known

$$x_i = A_i \cos 2\pi(vt + \phi) \qquad (5.59)$$

where v is the frequency of the mode, and ϕ is a phase factor.

The displacements, x_i, for each atom i can be found for all conceivable values of ε' where $\infty > \varepsilon' \geq \varepsilon_0$. The energy of an activated molecule, ε', is made up of a sum of all the energies, ε_j, for each normal mode of vibration

$$\varepsilon' = \sum_j \varepsilon_j. \qquad (5.60)$$

For each value of ε', all possible distributions of this energy, ε', between the normal modes are considered. All possible distributions of energy ε' are allowed subject only to the following two conditions

$$\infty > \varepsilon' \geq \varepsilon_0$$

$$\varepsilon' = \sum_j \varepsilon_j, \qquad (5.61)$$

so that ultimately the theory gives the amplitudes and hence the displacements for every conceivable allowable value of ε', and for every conceivable allowable distribution pertinent to the possible ε'.

The working equations converting spectroscopic data into a useful form for the Slater treatment are

$$A_{ij} = \alpha_{ij}' \sqrt{\varepsilon_j} \qquad (5.62)$$

$$x_{ij} = A_{ij} \cos 2\pi(v_j t + \phi_j) \qquad (5.63)$$

$$\begin{cases} \varepsilon' = \sum_j \varepsilon_j \\ \infty > \varepsilon' \geq \varepsilon_0 \end{cases} \tag{5.64}$$

where i stands for the atom involved, and j stands for the particular mode being studied.

The displacements x_{ij} are a periodic function of time Equation 5.59. If all these displacements were to be summed up for all modes, this would end up as a description of the overall changes of size and shape of a molecule as a function of time. This process is a summing of sine or cosine waves, and leads to "finding how the vibrations come into phase", which means finding when the total displacements gives the biggest maxima.

For instance, the summing up of three sine waves representing the periodic displacements of one atom in three normal modes of vibration is given in Figures 5.10, 5.11 and 5.12.

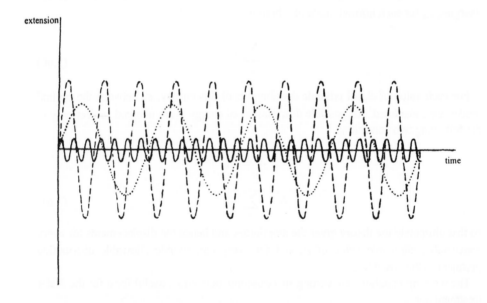

Figure 5.10 **Summing of sine waves showing the extensions produced by three normal modes individually.**

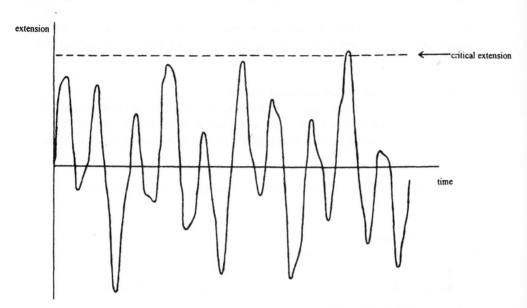

Figure 5.11 Summing of sine waves showing the resultant extension produced by three normal modes.

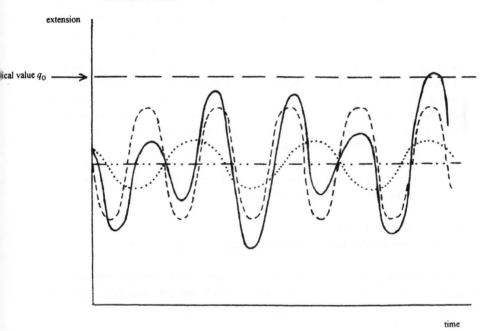

Figure 5.12 Summing of sine waves for the decomposition of N_2O, showing how the critical extension q_0, can be attained

_._._._	bending
....	symmetric stretch
------------	asymmetric stretch
_____	actual motion: sum of all three

Slater's theory does not require a complete summation over all atoms and all modes, but requires only a summation of all atoms and all modes which *make a contribution* to the critical coordinate defining reaction.

5.3.8 Significance of this for the Slater theory

Slater's theory looks at the structure of the reacting molecule, and then decides on a realistic critical configuration, and from a detailed study of this decides on what the critical coordinate has to be. The theory then assigns a critical value to this coordinate, which can be

(i) a bond length or lengths, (ii) a bond angle or angles, (iii) or any combination of bond length(s) or bond angle(s).

Once this decision is made as to what contributes to the critical coordinate, then all the atoms involved are identified, and from this all the normal modes which contribute to the critical coordinate are known. The displacements for each atom in the relevant modes can be found for all allowable values of ε', and all possible distributions of each value of ε' as described earlier.

The Slater theory then looks at the way in which these displacements contribute to the critical coordinate, and sums their contributions accordingly - in effect a summing of cosine and sine waves. This gives values of the critical coordinate as a periodic function of time. If the molecule has sufficient energy to be activated, that is, it has energy at least ε_o in its normal modes, and *if* this energy is suitably distributed among the normal modes of vibration which contribute to changes in the critical coordinate, then the critical coordinate defining reaction will exceed the critical value in a periodic manner. If, however, the activated molecule has energy at least ε' in its normal modes, but does not have this energy distributed in a suitable way among the normal modes which contribute to the critical coordinate, then reaction cannot occur because the critical coordinate will never reach its critical value.

A molecule can therefore be a potential reacting system if it has energy at least ε_o, but a very stringent restriction has been placed on when it can, or cannot, reach the critical configuration of the activated complex, and react.

5.3.9 The Slater calculation
Step 1. *This gives a value to the critical coordinate defining reaction*

The critical coordinate, q, is assessed from the geometry of the reactant molecule, and the normal modes contributing to the critical coordinate are identified. Using the procedure outlined early (Section 5.3.7) the important quantities

$$\alpha'_{ij} ; A_{ij} ; \varepsilon_j ; x_{ij}$$

are found, and the displacements are summed to calculate q as a periodic function of time

$$q = \left\{ \begin{array}{l} \text{sums and differences of the type} \\ \sum_j \alpha'_{ij} \sqrt{\varepsilon_j} \cos 2\pi (v_j t + \phi_j) \end{array} \right\} \qquad (5.65)$$

$$= \sum_j \left\{ \alpha_j \sqrt{\varepsilon_j} \cos 2\pi(v_j t + \phi_j) \right\} \tag{5.66}$$

where the sum is over all modes, j, which contribute to the critical coordinate. α_j replaces α'_{ij} as a shorthand way of talking about the amplitude factors and amplitudes of the modes j which contribute to the critical coordinate.

Step 2. *This gives the mathematical statement of the criterion for reaction*
The criterion of reaction is that reaction will occur if the critical coordinate, q, equals or exceeds a critical value q_0

$$q \geq q_0. \tag{5.67}$$

This will only happen if the normal modes happen to have a distribution of energy, ε', where $\infty > \varepsilon' \geq \varepsilon_0$, among them such that it is possible for $q \geq q_0$. Some situations turn up where the energy requirement is satisfied but the distribution of the energy is such that q cannot exceed q_0, and reaction will not occur.

This results in the mathematical description of conditions favourable to reaction as:

$$q \geq q_0$$

$$\sum_j \alpha_j \sqrt{\varepsilon_j} \cos 2\pi(v_j t + \phi_j) \geq q_0. \tag{5.68}$$

Figure 5.13 gives a pictorial description of the criterion for reaction.

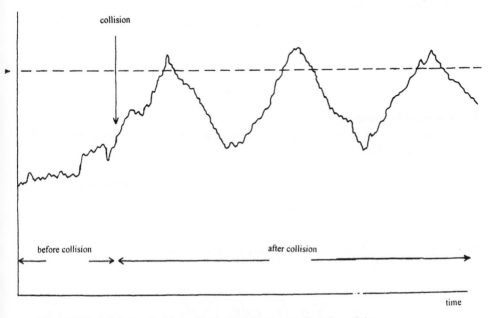

Figure 5.13 Attainment of the critical extension q_0 as a function of time.

Step 3. *This gives the mathematical statement of the criterion of reaction, and gives a calculation of ε_o*

The energy ε' which an activated molecule possesses is given by Equation 5.60. What is required is the value of ε' which is equal to ε_o, that is where

$$\varepsilon' = \varepsilon_0 = \sum_j \varepsilon_j$$

$$(5.69)$$

and which also allows q to become equal to q_o.
This turns out to be

$$\varepsilon_0 = \frac{q_0^2}{\sum_j \alpha_j^2}$$

$$(5.70)$$

where q_o is the value which the following summation must be able to reach

$$\sum_j \alpha_j \sqrt{\varepsilon_j} \cos 2\pi(v_j t + \phi_j).$$

Equation (5.70) allows ε_o to be calculated.

Step 5. *This gives a summary mathematical statement of the criteria for activation and reaction*

(i) *Criterion for activation*

For a molecule to be activated it must have energy ε' which is at least ε_o in its normal modes of vibration.

$$\varepsilon' \geq \varepsilon_0 = \frac{q_0^2}{\sum_j \alpha_j^2}.$$

$$(5.71)$$

(ii) *Criterion for reaction*

The molecule will not react unless the energy distribution is such that q can exceed q_o. This happens when

$\cos 2\pi(v_j t + \phi_j)$ is equal to unity for all j, so that Equation 5.68 becomes

$$\sum_j \alpha_j \sqrt{\varepsilon_j} \geq q_0.$$

Since $q_0 = \sqrt{\varepsilon_0} \sqrt{\sum_j \alpha_j^2}$ equation 5.70 rearranged (5.72)

then

$$\sum_j \alpha_j \sqrt{\varepsilon_j} \geq \sqrt{\varepsilon_0} \sqrt{\sum_j \alpha_j^2}.$$

$$(5.73)$$

Step 6. *This calculates the mean lifetime of the activated molecule*
The calculation gives the number of times per unit time that q reaches q_0, or greater. This quantity is the reciprocal of the mean lifetime of the activated molecule which is itself equal to the reciprocal of the rate constant, $k_2(\varepsilon')$, for reaction from the energy ε'.

The frequency with which q exceeds q_0 is a complex function of quantities already derived from the fundamental spectroscopic data.

$$L = \frac{\left(\sum_j a_j \sqrt{\varepsilon_j} - q_0 \right)^{\frac{1}{2}(n-1)}}{(2\pi)^{\frac{1}{2}(n-1)} C} \left(\frac{\sum_j a_j \sqrt{\varepsilon_j} v_j^2}{\prod_j a_j \sqrt{\varepsilon_j}} \right)^{\frac{1}{2}}$$

(5.74)

where n is the number of normal modes involved

if n is even: $C = \dfrac{n-1}{2} \times \dfrac{n-3}{2} \cdots \dfrac{3}{2} \times \dfrac{1}{2} \sqrt{\pi}$

(5.75)

if n is odd: $C = \left(\dfrac{n-1}{2} \right)!$

(5.76)

and $\prod_j$ means a product over all j

Figure 5.13 shows a plot of the value of q with time. Before collision the molecule has not enough energy to be an activated molecule, and so q never exceeds q_0. After collision the molecule is activated, and has the energy distributed in the appropriate normal modes which allow reaction, and q can exceed q_0.

From this graph the average number of times per unit time (the frequency of crossings, L) that q can exceed q_0 can be found, and L quantified. The time intervals between successive crossings, that is between times at which $q \geq q_0$ gives the mean lifetime of the activated molecules which have the appropriate energy distributions. The reciprocal of this lifetime gives $k_2(\varepsilon')$

$$L = \frac{1}{\tau} = k_2(\varepsilon').$$

(5.77)

Constructing the graph is a very difficult mathematical procedure, especially for polyatomic molecules where the number of atoms and normal modes is large.

Step 7. *This calculates k_∞ and k_{obs}^{1st}*

$$k_\infty = \sum_{\varepsilon'} \frac{k_1(\varepsilon')k_2(\varepsilon')}{k_{-1}}.$$

(5.78)

The value of $k_2(\varepsilon')$ is found in Step 6 and k_{-1} is assumed to be a constant equal to λZ, which leaves $k_1(\varepsilon')$ to be found. As previously, $k_1(\varepsilon')$ is a rate constant for accumulation of energy into the total number of normal modes, 1, 2 ... n, where the energy in mode 1 is in the range $\varepsilon_1 + \delta\varepsilon$, mode 2 is in the range $\varepsilon_2 + \delta\varepsilon$ ...   mode n is in the range $\varepsilon_n + \delta\varepsilon$, and

$$\varepsilon' = \varepsilon_1 + \varepsilon_2 + \ldots + \varepsilon_n. \tag{5.79}$$

The Maxwell-Boltzmann distribution then leads to $k_1(\varepsilon')/k_{-1}$.

$$\frac{k_1(\varepsilon')}{k_{-1}} = \frac{\exp\left(-\dfrac{\varepsilon'}{kT}\right)\delta\varepsilon_1 \delta\varepsilon_2 \cdots \delta\varepsilon_n}{(kT)^n}. \tag{5.80}$$

Insertion of this expression for $k_1(\varepsilon)/k_{-1}$ and that for $k_2(\varepsilon') = L$ (Equation 5.74), into Equation 4.39) leads to the value of k_∞.

$$k_\infty = \iint_{\varepsilon_1+\varepsilon_2+\cdots\varepsilon_n \geq \varepsilon_0} \cdots \int \exp\left(-\frac{\varepsilon'}{kT}\right) \frac{L d\varepsilon_1 d\varepsilon_2 \cdots d\varepsilon_n}{(kT)^n}. \tag{5.81}$$

The total energy in modes 1, 2 ... n is given by

$$\varepsilon' = \varepsilon_1 + \varepsilon_2 + \ldots + \varepsilon_n. \tag{5.82}$$

and the lower limit for this is ε_0.

Once the complex expression (5.74) for L is explicitly quoted, and the following condition is included

$$q_0 = \sqrt{\varepsilon_0} \sqrt{\sum_j \alpha_j^2} \tag{5.72}$$

Equation 5.78 leads eventually to

$$k_\infty = v \exp\left(-\frac{\varepsilon_0}{kT}\right) \tag{5.83}$$

where

$$v = \frac{\left(\sum\limits_j \alpha_j^2 v_j^2\right)^{\frac{1}{2}}}{\left(\sum\limits_j \alpha_j^2\right)^{\frac{1}{2}}}. \tag{5.84}$$

This equation for k_∞ is exactly of the Arrhenius form, and the theoretical ε_0 can be identified with the high pressure experimental activation energy, and v with the corresponding pre-exponential factor.

Calculation of k_{obs}^{1st} makes use of equation

$$k_{obs}^{1st} = \sum_{\varepsilon'} \frac{k_2(\varepsilon')k_1(\varepsilon')/k_{-1}}{1+k_2(\varepsilon')/k_{-1}[A]} \tag{5.85}$$

giving

$$k_{obs}^{1st} = \iint_{\varepsilon'+\varepsilon_2+\cdots\varepsilon_{n\geq\varepsilon}} \cdots\int \frac{L\exp\left(-\dfrac{\varepsilon'}{kT}\right)d\varepsilon_1 d\varepsilon_2 \cdots d\varepsilon_n}{(kT)^n(1+L/k_{-1}[A])} \tag{5.86}$$

Integration of this expression is carried out by substituting Equation 5.74 for L, and is subject to the following two conditions.

(a)
$$\varepsilon' \geq \varepsilon_0 = \sum_{\varepsilon_j} \frac{q_0^2}{\alpha_j^2}, \tag{5.87}$$

(b)
$$\sum_j \alpha_j \sqrt{\varepsilon_j} \geq q_0. \tag{5.88}$$

(a) being the criterion for activation, and (b) being the criterion for reaction.

If the integral is evaluated, then a theoretical fall-off curve for k_{obs}^{1st} can be found. The calculations are extremely difficult and lengthy even using a computer. It is partly because of this that only a few reactions have been analysed using Slater's theory. However, the main reason is that the Marcus theory and its extensions have largely superseded Slater's theory. The calculations involved in the RRKM theory, though difficult and lengthy, are still easier to carry through than those for this classical harmonic Slater theory. The quantum version, and the version which allows energy flow are even more difficult, and this explains why not much has been done on developing a quantum non-harmonic (allowing energy flow) version.

5.3.10 Final comments on Slater's theory
1. The theory requires a complete vibrational analysis of the reacting molecule, and this data may not always be available. Moreover, it assumes that the behaviour of a molecule on the point of dissociation can be suitably estimated from the properties of a molecule in a non-excited state.
2. The vibrational analysis into normal modes of vibration requires the assumption of harmonic oscillators. Experiment makes it absolutely clear that vibrations are anharmonic, especially at high vibrational levels. Anharmonicity must be allowed for, but this would allow a flow of energy between the normal modes of vibration - a situation not allowed for in the version of Slater's theory given here. Such an energy flow is more likely to happen at pressures where there is a long time interval between collisions, and the lifetime of an activated molecule be relatively long. If an energy flow is allowed, the Slater theory is vastly improved and gives a much better fit at low pressures. Making this modification brings the theory more on a par with Marcus' theory.

3. As with most theories, the quantum version is more satisfactory than the classical version. However, the quantum version is not so fully developed as the classical version, and is considerably more complex.

4. The theory is poor for small molecules having few normal modes of vibration.

5. The theory imposes a very stringent requirement on the criterion of reaction, and is much more restrictive than any of the other theories considered.

6. *The beauty of Slater's theory is that it gives a very clear description of the process of reaction which is easy to visualise*, that is "the molecule splits up if some bond(s), bond angle(s), or a combination of both is extended too far". This is in contrast to Marcus' theory and its extensions where thinking about the model included several very complex concepts.

6

Potential Energy Surfaces

The potential energy surface plays a key role in understanding what happens during chemical reaction, and it is vital for any theoretical description of the rate of reaction. Recent developments in both theoretical and experimental kinetics are intimately bound up with modern calculations of potential energy surfaces, and more reliance can now be placed on these accurate surfaces, and on the details which appear on them.

Modern, fast, high capacity computers have been vital for the development of molecular dynamics where movement across the potential energy surface is studied as a function of the reactant's initial conditions. These trajectory calculations show how the topography of the surface affects the dynamics and energetics of movement over the surface. The results are then checked by molecular beam and chemiluminescence experiments. The reverse procedure is often used, where the molecular beam experimental results are used to infer the details on the surface.

Although basically an exercise in quantum mechanics, the calculation of the potential energy surface for reaction is an integral part of theoretical kinetics. There are many ways to calculate these surfaces, but essentially they break down into two techniques; *ab initio* methods and *semi-empirical* methods.

6.1 Quantum ab initio calculations
Up to the recent past, accurate potential energy surfaces could only be calculated by ab initio methods for relatively small molecules containing light atoms. For larger molecules the calculations are more difficult, and more demanding of computer time and capacity, but modern computing technology is rapidly extending these calculations to large molecules.

Ab initio methods use particular models for the electronic wave function, and the success of the calculation depends on the choice of the model. Two methods are used, one based on the original London, Eyring and Polanyi method, and the other on the variation technique of quantum mechanics.

6.1.1 Original London, Eyring, Polanyi method
The total energy, E, of a system of electrons and nuclei is given approximately in terms of coulomb and exchange integrals.

For the very simple case of reaction between

$$X + YZ \quad \rightarrow \quad XY + Z$$

where X, Y and Z are atoms with one electron each, and the configuration of the "reaction entity" X...Y...Z is as in Figure 6.1, the problem reduces to a three electron problem where

$$E = A + B + C \pm \left\{ \frac{1}{2} \left[(\alpha - \beta)^2 + (\beta - \gamma)^2 + (\gamma - \alpha)^2 \right] \right\}^{\frac{1}{2}} \quad (6.1)$$

A, B, C are coulomb integrals for each electron pair, Figure 6.1

α, β, γ are the corresponding exchange integrals,

the zero of energy is the separated atoms at rest, and

$E =$ potential energy of electron-electron and electron-nuclear interactions

       + potential energy of nuclear interactions

             + kinetic energy of the electrons.

The kinetic energy of the nuclei is *not* included, the Born-Oppenheimer approximation, and it is conventional to call E the "potential energy of the system" *even though* it *does include* the kinetic energy of the electrons.

$$\longleftarrow r_1 \longrightarrow \longleftarrow r_2 \longrightarrow$$
$$X \cdots\cdots Y \cdots\cdots Z$$
$$\longleftarrow\!\longrightarrow r_3 \longrightarrow$$

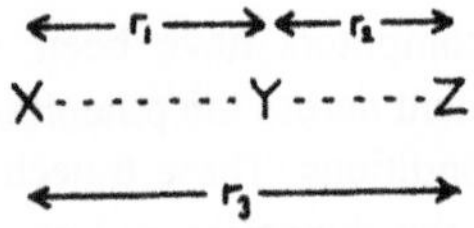

Figure 6.1   **Simplified configurations for reaction: collinear approach and recession. Each of the integrals A, B, C, α, β, γ is an integral for a 2-electron system, and is associated with two of the nuclei: A (which depends on r_1) is the coulomb integral for the 2-electron system which would remain if nucleus Z and one electron were removed, and α is the corresponding exchange integral. The integrals B and β correspond to the system with nucleus X and one electron removed, and C and γ to the system with nucleus Y and one electron removed.**

A three dimensional surface can only be generated if X, Y, Z are all atoms, and if the activated complex (X...Y...Z) is linear. Values of the potential energy, E, are found by evaluating the coulomb and exchange energies for various arrangements of the atoms, Figure 6.1, and lead to a potential energy surface for the reaction.

Equation 6.1 can be rewritten to correspond to different reactions. Full quantum mechanical calculations of the integrals have been carried out for small molecules, the extension to ever larger molecules depends on the capacity of the computer used. Unfortunately, no matter how accurate the calculations are, they are limited by the very serious approximations inherent in the derivation of the original London, Eyring, Polanyi equation.

6.1.2 Variational methods

In these techniques the wave function is varied systematically, and the wave function giving the lowest energy is taken to be the best approximation. With modern computers trial wave functions with many terms and many parameters can be varied, and these give reliable surfaces and activation energies. Extension to more complex molecules is limited by the speed and memory of the computer used.

6.2 Semi-empirical calculations

These calculations aim at obtaining a fit with experiment, rather than to generate independent trial surfaces, and use experimental data in conjunction with quantum calculations. Unfortunately, large approximations often have to be made to obtain good fits.

One method uses the Morse equation along with spectroscopic data to calculate values of ε_1, ε_2, ε_3 as a function of distance where

$$\varepsilon_1 = A + \alpha \qquad \text{for various } r_1$$
$$\varepsilon_2 = B + \beta \qquad \text{for various } r_2$$
$$\varepsilon_3 = C + \gamma \qquad \text{for various } r_3$$

where r_1, r_2 and r_3 are defined as in Figure 6.1.

Separate values of A, B, C, α, β, γ are needed before fitting values into Equation 6.1 is possible. The original procedure assumed that the coulomb energy is a constant fraction of the total energy. Values of this fraction can be assumed, and that giving the best fit can be found. This is a large approximation, and can lead to spurious features, such as potential wells, on the surface.

It is possible to allow the fraction, coulomb energy divided by total energy, to vary with internuclear distance. This appears at first sight to be an improvement on the assumption of a constant fraction, but any advantages at this stage may well be balanced by inaccuracies of the assumed distance dependence of the ratio. Values of the separate integrals can then be fed into the London, Eyring, Polanyi equation.

Another semi-empirical approach involves bond energies and bond orders. The calculation is restricted to potential energies along the reaction coordinate where the X - Y bond is being formed and the Y - Z bond is breaking. Only partial bonds X...Y...Z are present as the "unit" XYZ moves along the reaction coordinate.

Three semi-empirical relationships are used:
(1) the dependence of the bond order, n, on the internuclear distance can be found if the single bond length for the pair of atoms is known,
(2) the bond energy of the partial bond and the bond order are related to each other, provided the single bond energy is known
(3) the sum of the bond orders for all the bonds equals unity.
     Changes in the energies of the bonds involved can then be found as a function of distance along the reaction path, and this gives the barrier height and the critical configuration of the activated complex.

6.3 Potential energy contour diagrams and profiles

For the general reaction

$$A + BC \quad \rightarrow \quad AB + C$$

the reaction unit is (A...B...C) and the progress of the unit A...B...C along the reaction coordinate can be summarised in terms of the relative values of r_{AB} and r_{BC}, which can then be correlated with the potential energy surface. This is given here, Figures 6.2a, 6.2b, in terms of the 2-dimensional contour and profile diagrams appropriate to the three atom linear activated complex and the minimum energy path. The features discussed below are common to both the 2-dimensional contour and profile, and also to the corresponding *n*-dimensional diagrams relevant to the general case where A, B and C can be atoms or molecules.

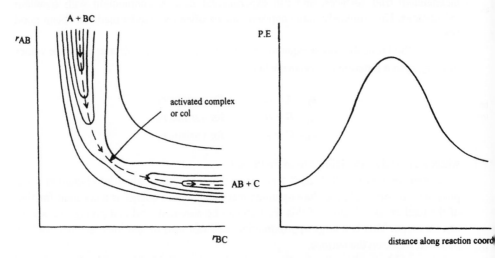

Figure 6.2a Potential energy contour diagram showing the reaction coordinate as the minimum energy path, and the position of the critical configuration defining the activated complex.

Figure 6.2b The potential energy profile derived from the minimum energy path, the reaction coordinate, in Figure 6.2a

At the entrance to the entrance valley r_{AB} is large and decreases towards the col. At the entrance to the exit valley r_{BC} is large and decreases towards the col. The position of the col can be described in terms of the relative magnitudes of the distances r_{AB} and r_{BC}.

The overall energy change on reaction can be classified as exothermic, endothermic or thermoneutral, Figure 6.3. Potential energy surfaces will be discussed in turn for each type of reaction.

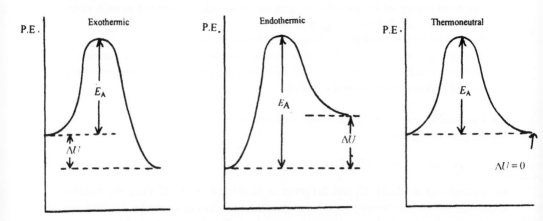

distance along reaction coordinate

Figure 6.3 Potential energy profiles for exothermic, endothermic and thermoneutral reactions.

6.3.1 Types of potential energy barriers

Four types of barrier can be considered, and these are illustrated in the various diagrams of Figure 6.4.

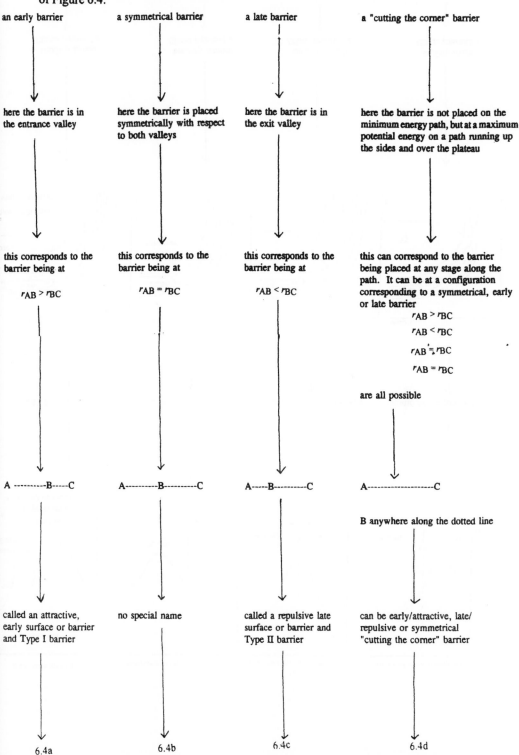

an early barrier	a symmetrical barrier	a late barrier	a "cutting the corner" barrier
here the barrier is in the entrance valley	here the barrier is placed symmetrically with respect to both valleys	here the barrier is in the exit valley	here the barrier is not placed on the minimum energy path, but at a maximum potential energy on a path running up the sides and over the plateau
this corresponds to the barrier being at	this corresponds to the barrier being at	this corresponds to the barrier being at	this can correspond to the barrier being placed at any stage along the path. It can be at a configuration corresponding to a symmetrical, early or late barrier
$r_{AB} > r_{BC}$	$r_{AB} = r_{BC}$	$r_{AB} < r_{BC}$	$r_{AB} > r_{BC}$ $r_{AB} < r_{BC}$ $r_{AB} = r_{BC}$ $r_{AB} = r_{BC}$ are all possible
A ---------B-----C	A----------B----------C	A----B----------C	A--------------------C B anywhere along the dotted line
called an attractive, early surface or barrier and Type I barrier	no special name	called a repulsive late surface or barrier and Type II barrier	can be early/attractive, late/repulsive or symmetrical "cutting the corner" barrier
6.4a	6.4b	6.4c	6.4d

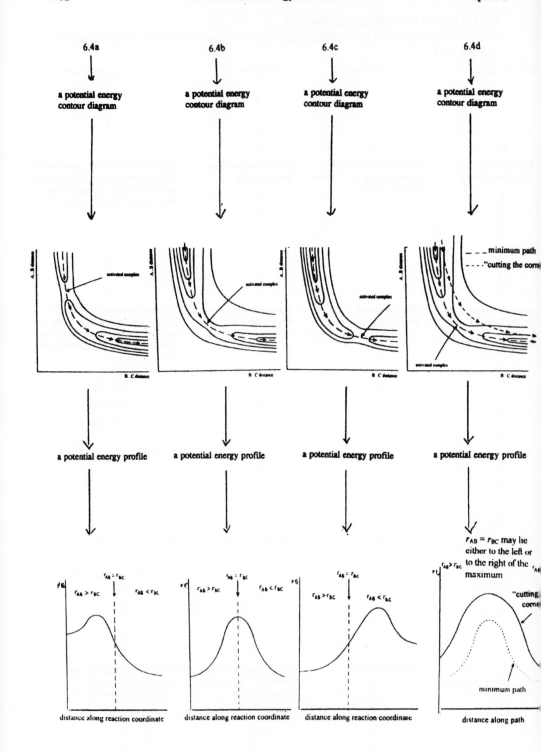

6.3.2 Features of the potential energy surface

The potential energy barrier can be symmetrically placed, or it can appear in the exit or entrance valleys. By a suitable choice of coordinates these features can be altered systematically, for instance the barrier could be moved from the entrance to the exit valley, or a potential well could be introduced. Altering the potential energy surface in such ways is now a fundamental and growing area, especially as molecular dynamics calculations for movement across the surface now give specific predictions about the properties of the surface, and these predictions can be accurately tested by modern molecular beam and chemiluminescence experiments, see Chapter 8. The reverse procedure is also used. Results from molecular beam experiments can be used to construct the potential energy surface and to add empirically determined features. For instance, molecular beam experiments can distinguish between "direct" reaction or reaction via "a collision complex". If there is evidence for a collision complex this feature can be added to the calculated surface. Other experiments give indications about the possible steepness of the barrier, or curvature of the reaction coordinate. Molecular dynamics studies, described below, can then compare surfaces with and without such added features. This is particularly useful for more complex reactions where the totally quantum calculation is difficult, and for giving detailed information about the surface around the critical configuration for which the quantum calculation is at its least accurate.

Every reaction and every potential energy surface is unique, but nonetheless there are general features of surfaces for which molecular dynamics can make predictions and molecular beams can test, and these can be discussed without reference to specific reactions.

6.4 Molecular dynamics

Molecular dynamics techniques calculate trajectories across potential energy surfaces. Each trajectory describes the movement of the "reaction unit" (X...Y...Z) across the surface. These surfaces can be pure quantum surfaces for simple reactions, and may have, or not have, added features, but more often semi-empirical or experimentally modified potential energy surfaces are used.

Reaction on a potential energy surface is described in terms of the variation of interatomic distances with time. The forces of interaction at each position can be calculated once the potential energy is known, and the movement of (X...Y...Z) under the influence of these forces can be calculated. This movement should ideally be described in quantum terms, but such calculations are notoriously difficult and classical equations of motion are solved instead, and give a classical trajectory. Fortunately, the likely errors in the classical treatment are predictable, and corrections can be made. The calculations become more reliable with increase in atomic mass, with increasing energy for each degree of freedom considered, and also by virtue of the averaging procedures inherent in the procedure. The calculations are done by computer, and not surprisingly more detailed calculations result from faster high memory computers.

Most of the really important requirements and consequences of reaction can be studied in this way, and then compared with experiment. For instance, the chance of reaction and the cross sectional area for reaction can be found for a range of impact parameters and angles of approach, Chapter 7. The energetics of reaction can also be studied systematically. Molecular dynamics shows how, in turn, translational, rotational and vibrational energy in the reactants affect the trajectory, and in addition allows calculation of the likely threshold energies for reaction for each type of energy. Likewise, the effect of the type of potential energy surface on the distribution of energy in the products can be predicted, and this then can lead to a discussion of the redistribution of energy on reaction.

6.5 Trajectory calculations

A large number of sets of initial conditions are chosen, and these are randomised by a Monte Carlo averaging procedure. These, and the details of the particular potential energy surface under scrutiny are fed into the computer, the classical equations of motion are solved, and a trajectory results for each set of initial conditions. The whole procedure is then repeated for other initial conditions, and finally the whole procedure is repeated for different types of potential energy surfaces.

Standard molecular dynamics calculations show that there are both reactive and non-reactive collisions. The reactive ones obviously make it to the products' configuration, and the non-reactive do not. These can be distinguished by comparing plots of changing interatomic distances with time for both category of collision where, in Figure 6.5

$$A + BC \quad \rightarrow \quad AB + C \qquad \text{is a reactive collision}$$

and in Figure 6.6

$$A + BC \quad \rightarrow \quad A + BC \qquad \text{is a non-reactive collision}$$

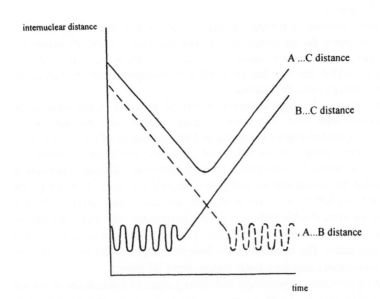

Figure 6.5   A trajectory diagram showing a reactive collision for reaction
                  A + BC → AB + C

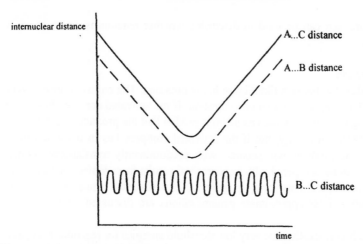

Figure 6.6 A trajectory diagram showing a non-reactive energy transfer collison
         $A + BC \rightarrow A + BC$

The plots show the variation of the BC, AB and AC distances with time

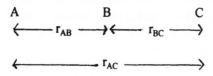

The results are summarised in Table 6.1

Table 6.1   Reactive and non-reactive collisions

(a) Reactive collision	(b) Non-reactive collision
(i) Before collision occurs A approaches BC, and the AB distance decreases	(i) Throughout the trajectory A approaches BC, and the AB distance decreases until A hits BC
(ii) The average BC distance stays approximately the same showing, in effect, the vibration of BC	(ii) No reaction occurs, and the AB distance increases as the two part company
(iii) The AC distance decreases	(iii) The average BC distance stays approximately the same, showing the vibration of BC
(iv) After reaction occurs the average AB distance stays approximately the same, showing vibration of AB. BC distance increases, AC distance increases	(iv) The AC distance first decreases and then increases as A approaches BC, collides and parts company

This sort of approach can be used to determine whether reaction is direct or via a collision complex.

6.5.1 Direct Reaction

If the reaction is direct a diagram like Figure 6.5 is obtained. The collision time is very short and the products separate within one vibration. If the activated complex lies in the entrance valley, Figure 6.4a, then the calculations predict that the products separate with predominantly vibrational energy, but if the activated complex lies in the exit valley, Figure 6.4b, then the products will separate with predominantly translational energy. These predictions can be compared with experimental results from molecular beam and chemiluminescence experiments, and, in general, experimental results confirm the trajectory predictions. Less approximate generalisations are discussed later (Sections 6.10 - 6.14).

If the reaction is highly exothermic very low threshold energies for reaction result, and the chance of reaction is almost independent of translational energy.

6.5.2 Reaction with a collision complex and well

This gives trajectories as in Figure 6.7 suggestive of a long-lived collision complex which persists for many vibrations, or even several rotations. The lifetime of the collision complex is long enough to allow equilibration between vibration and translation, so that the products separate with comparable amounts of both. Again these conclusions have been confirmed by molecular beams.

A collision complex will show up as a potential energy well in the surface, Figure 6.8a, and as a minimum in the profile, Figure 6.8b. If reaction is direct there will be no such feature. Molecular beam experiments can give conclusive evidence as to whether reaction is "direct" or via a "collision complex".

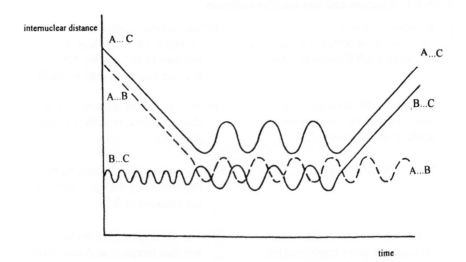

Figure 6.7   A trajectory diagram for a reactive collision giving a complex followed by reaction A + BC → [A-B-C] → AB + C

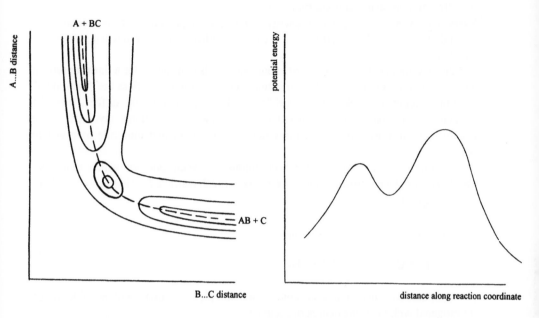

Figure 6.8a A potential energy contour
diagram for a reaction involving a
collision complex which shows up
as a potential energy well.

Figure 6.8b The potential energy profile
for a reaction proceeding via a collision
complex.

6.6 Testing of predictions of transition state theory

Results from trajectory calculations show that transition state theory is generally more
satisfactory for symmetrical barriers, and much less so for asymmetric ones.

For instance, the potential energy surface for the endothermic reaction

$$Br \cdot + H_2 \quad \rightarrow \quad HBr + H \cdot$$

has added features from molecular beam experiments which suggest that there is a high
late energy barrier for reaction in the forward direction, and a low early barrier for the
reverse reaction.

In the *forward direction*, good agreement is generally found for low energies, but at
higher energies the agreement is poorer. Trajectory calculations show that at the higher
energies the transition state quite often is never reached, although units (Br...H...H)
which do attain the transition state go straight to products. At higher energies there is a
fair chance that units (Br...H...H) could be bounced back by the inner walls and retreat
back down the reactant valley without reaching the transition state.

For the *reverse reaction* good agreement is found between transition state theory and
trajectory calculations at low energies, but again discrepancies appear at high energies.
These discrepancies are explained by assuming that some units (Br...H...H) which
actually do reach the transition state are bounced back by the inner walls at the transition
state, and return to the reactant valley instead of running into the product valley.

6.7 Reaction in terms of mass type

Every reaction and every potential energy surface are unique, but there are, nonetheless, general features of the surfaces which can be studied without reference to specific reactions.

If the "traditional" type of potential energy surface is considered for a linear approach and recession, Figure 6.1, and the minimum energy path which defines the "traditional" reaction coordinate is taken to be the "most likely" path over the surface, then the conclusions turn out to be dependent on the masses of the atoms involved in the reaction, or more exactly on the mass ratios of the atom and molecule involved in reaction.

Reactions whose surfaces have been studied in detail, and for which molecular trajectories can be constructed are at present very much restricted to simple ones such as

$$K\cdot + HCl \quad \rightarrow \quad KCl + H\cdot$$

or the slightly more complex ones such as

$$K\cdot + CH_3I \quad \rightarrow \quad KI + CH_3\cdot$$

As computational techniques improve more complex reactions will be able to be investigated with the same degree of accuracy.

The reaction scheme is classified into categories defined in terms of the masses, and mass ratios of the atoms involved. Heavy atoms are labelled H and light atoms are labelled L, and all possible combinations of H and L forming a molecule are defined.

6.7.1 Consider atom H approaching the various molecules

1. $H + HH \quad \rightarrow \quad HH + H$

typified by $K\cdot + Br_2 \quad \rightarrow \quad KBr + Br\cdot$

2. $H + HL \quad \rightarrow \quad HH + L$

here the H atom approaches the H end of the molecule HL
typified by $K\cdot + HCl \quad \rightarrow \quad KCl + H\cdot$

3. $H + LH \quad \rightarrow \quad HL + H$

here the H atom approaches the L end of the molecule LH
typified by $Cl\cdot + HBr \quad \rightarrow \quad HCl + Br\cdot$

4. $H + LL \quad \rightarrow \quad HL + L$

typified by $Br\cdot + H_2 \quad \rightarrow \quad HBr + H\cdot$

The reverse of these reactions are

1′ $H + HH \quad \rightarrow \quad HH + H$
2′ $L + HH \quad \rightarrow \quad LH + H$

3' $HL + H \quad \rightarrow \quad H + LH$

Here the H atom approaches the L end of the molecule LH

4' $HL + L \quad \rightarrow \quad H + LL$

Here the L atom approaches the L end of the molecule HL.

6.7.2 Consider atom L approaching the various molecules

5        $L + LL \quad \rightarrow \quad LL + L$
typified by $H \cdot + D_2 \quad \rightarrow \quad HD + D \cdot$

6        $L + HL \quad \rightarrow \quad LH + L$

Here the L atom approaches the H end of molecule HL
typified by $D \cdot + HBr \rightarrow \quad DBr + H \cdot$

7        $L + LH \quad \rightarrow \quad LL + H$

Here the L atom approaches the L end of the molecule LH
typified by $H \cdot + HI \quad \rightarrow \quad H_2 + I \cdot$

8        $L + HH \quad \rightarrow \quad LH + H$
typified by $H \cdot + Br_2 \quad \rightarrow \quad HBr + Br \cdot$

The reverse of these reactions are

5' $L + LL \quad \rightarrow \quad LL + L$
6' $L + HL \quad \rightarrow \quad L + HL$

Here the atom L approaches the H atom of the molecule LH

7' $H + LL \quad \rightarrow \quad HL + L$
8' $LH + H \quad \rightarrow \quad L + HH$

Here the atom H approaches the H atom of the molecule LH

6.8 Exothermic and endothermic reactions
Endothermic reactions are the exact reverse of **exo**thermic reactions, Figure 6.3 so that, for example

the **endo**thermic reaction $H + HL \quad \rightarrow \quad HH + L$ is the reverse of
the **exo**thermic reaction $L + HH \quad \rightarrow \quad LH + H$ and
the **endo**thermic reaction $L + HH \quad \rightarrow \quad LH + H$ is the reverse of
the **exo**thermic reaction $H + HL \quad \rightarrow \quad HH + L$

and the potential energy surface contour diagrams and potential energy surface of the *endo*thermic reaction are the exact reverse of the *exo*thermic reaction.

In these reactions, that is 1 to 8, the exothermic reaction will involve a light/heavy atom as the reactant where the reverse endothermic reaction involves the heavy/light atom. Care must be taken to watch out for such switches in mass type.

The arguments and conclusions for the *endo*thermic reaction are the reverse of those for the *exo*thermic reaction, so that the **attractive** surface of an *exo*thermic reaction becomes a **repulsive** surface for the *endo*thermic reaction, and the **repulsive** surface of an *exo*thermic reaction becomes the **attractive** surface for the reverse *endo*thermic reaction.

For the endothermic reaction, the reactants must have a high energy to allow the "reaction unit" to reach the critical configuration, Figure 6.9. For the reverse exothermic reaction the energy requirement is not so drastic. This can be put equivalently in terms of the vertical distances marked on Figure 6.9.

(a)   The energy requirement for the forward endothermic reaction

=   AB

=   CD + AC

(b)   The energy requirement for the reverse exothermic reaction

=   CD

=   AB - AC

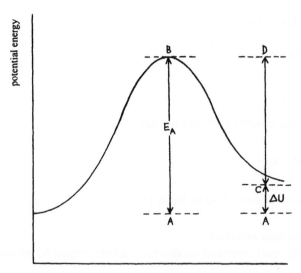

distance along reaction coordinate

Figure 6.9 Potential energy profile for an endothermic reaction.

6.9   More detailed aspects of the features of different types of potential energy barriers

Trajectory calculations describe the movement of the unit (A...B...C) across the potential energy surface. Trajectories will describe reaction if they start from the reactant configuration, A + BC, and move to the product configuration, AB + C, by any continuous route across the potential energy surface. The reaction coordinate up the entrance valley, over the col, and down into the exit valley is the minimum energy pathway and is generally the most probable route. However, other trajectories do contribute to the rate, and these paths are displaced from the reaction coordinate. The curvature of a path is a measure of its displacement from the reaction coordinate, Chapter 3, Figure 3.10 and the curvature of these trajectories can be affected by the mass type. Furthermore, the trajectories which result are also affected by whether the barrier is sudden or gradual, Section 6.20 following.

All four features, type of barrier, curvature and suddenness of the barrier and mass type are important in discussing the likely behaviour of the unit A...B...C as it moves over any given surface.

Trajectory calculations have been carried out for various surfaces, and for different initial conditions. Considerable attention has been given to studying the effects on the trajectories of translational, rotational and vibrational energy in the reactants, and to finding the likely distribution of energy disposal in the products. These features can then be correlated with different types of surfaces, and the predictions and correlations can then be tested by molecular beam and chemiluminescence experiments, Chapter 8. For a given surface, the conclusions of the trajectory calculations depend on the masses of the atoms in the reaction.

General features of attractive early barriers and repulsive late barriers can be inferred from these trajectory calculations. It must be stressed however, that these are general features only, and that additional factors may have to be taken account of for some specific reaction types.

The following general features are pertinent to the mass types described in Section 6.7. However, there are important exceptions which will be discussed later.

6.10   General features of attractive potential energy surfaces for exothermic reactions

The general reaction

$$A + BC \quad \rightarrow \quad AB + C$$

can be discussed in terms of

(i)   the reaction unit (A...B...C)
(ii)   the bonds A-B and B-C
(iii)   the distances r_{AB} and r_{BC}

On attractive surfaces, Figure 6.4a the unit (A...B...C) reaches the barrier or col before the bond B-C has altered much, and starts to move into the exit valley with A still far from BC, but with the BC distance still close to its equilibrium internuclear distance. The barrier is, therefore, placed in the entrance valley. The col, therefore, corresponds to the configuration

A ----------B-----C $r_{AB} \gg r_{BC}$

In the exit valley C separates from B when A is still far from B, and even when C has separated from B by a very considerable extent, the AB distance remains large compared to its equilibrium internuclear distance. And so the reaction energy is released as the new bond is forming. Repulsion between B and C cause B and C to recoil from each other as they separate so that B is propelled towards the approaching A *as the AB bond is forming*.

$$\begin{array}{ccc} \rightarrow & \leftarrow & \rightarrow \\ A \text{------------------} & B \text{--------} & C \end{array}$$

The motion set up can be loosely thought of as the extension of the bond B-C during vibration, and the inward approach of A and B towards each other in the vibration of the bond A-B. And so separation of BC causes vibrations to be set up in the A-B bond as it is forming. The amplitude of the vibration is large, and since the vibrational energy is proportional to the amplitude^2, then AB will be considerably vibrationally excited. The net result is that the fraction of vibrational energy in the products is much higher than the fraction of translational energy.

Attractive surfaces with early barriers are associated with predominantly vibrational energy in the products for such mass types as

(i)   the *exo*thermic reactions $L + HH \rightarrow LH + H$
                                      $H + HL \rightarrow HH + L$
(ii)   the **endo**thermic reactions $H + HL \rightarrow HH + L$

Attractive surfaces with early barriers are enhanced by high translational energy in the reactants, with vibrational energy playing a minor role. This is relevant to mass types, such as (i) and (ii) immediately above.

These conclusions are summarised in Table 6.2

Vibrational energy is less important for surmounting the attractive early barrier because high vibrational energy would encourage the "reaction unit" to hit onto the barrier wall perpendicular to the reaction coordinate at the end of the straight run up the entrance valley, from which it would be bounced back down the entrance valley.

Table 6.2 Approximate generalisations for energy requirements for reactants, and for energy disposal in products in terms of the type of potential energy surface and mass ratios

	Exothermic Reactions $\Delta H < 0$		Endothermic Reactions $\Delta H > 0$	
	$L + HH$	$H + HL$	$L + HH$	$H + HL$
Attractive surface with early barrier	translation effective; predominantly vibration in products	translation effective; predominantly vibration in products	translation and vibration effective; translation and vibration in products	translation effective; predominantly vibration in products
Repulsive surface with late barrier	vibration effective; predominantly translation in products	translation and vibration effective; translation and vibration in products	vibration effective; predominantly translation in products	vibration effective; predominantly translation in products

(a) $L + HL$, $L + LH$ similar to $L + HH$, but less extreme with order being $L + HH$, $L + HL$, $L + LH$

(b) Other combinations $H + LH$, $H + LL$, $H + HH$ and $L + LL$ are similar to $H + HL$, but less extreme

Trajectory calculations predict that high translational energy in the reactants enhances reaction, while high vibrational energy has little or no effect when the vibrational energy is greater than the vibrational threshold value. This effect is called "translational enhancement" or "translational promotion". At high translational energies the excess translational energy above the translational threshold value appears as translational energy in the products.

These conclusions derive from theoretical calculations, and experimental confirmation is essential. State to state molecular beams can do this admirably, and can also give further details.

Both theory and experiment show that for both *endo*thermic and *exo*thermic attractive surfaces of the mass type

(i) exothermic reactions: $L + HH$ $\rightarrow$ $LH + H$
$\phantom{\text{(i) exothermic reactions: }}H + HL$ $\rightarrow$ $HH + L$

(ii) endothermic reactions: $H + HL$ $\rightarrow$ $HH + L$

the following generalisations can be made.

1. Translational energy enhances reaction.
2. Vibrational energy is dominant in the products.
3. Selective enhancement by translational energy is easiest when there is a ***straight*** run up the entrance valley to the critical configuration. Vibrational energy would cause the reaction unit (A...B...C) to hit into the sides of the walls and be reflected back.

4. Product energy is predominantly vibrational when there is a *curved* route from the entrance of the exit valley up to the critical configuration of the activated complex at the col.
5. Determination of which type of energy gives selective enhancement can be phrased alternatively in terms of the momentum of the unit (A...B...C) along the reaction coordinate. When movement into the entrance valley has momentum parallel to the entrance valley then the unit (A...B...C) will be carried over the barrier. Parallel momentum is equivalent to translation, and so for an attractive barrier translational enhancement will occur.

Molecular beam experiments confirm these conclusions, and extend the results.

Attractive surfaces are normally associated with the forward scattering of stripping reactions where A approaches BC and from a distance, grabs hold of B and continues on undeflected, Chapter 10. Stripping reactions are, therefore, normally accompanied by large impact parameters and large cross sections. Here A is at large distances from B when BC splits, and B and C separate under the attractive forces between A and B. These intermolecular attractive forces are strong since they allow reaction of A with BC even at very large impact parameters. This holds for the reactions

(i)   exothermic $H + HL \rightarrow HH + L$
(ii) endothermic $H + HL \rightarrow HH + L$

However, when reaction involves the mass type

$$L + HH \rightarrow LH + H$$

the opposite happens, and even very attractive surfaces show backward scattering. This is explicable in terms of the high kinetic energy of separation.

6.11 General features of repulsive energy surfaces for exothermic reactions
Unlike the attractive surface, discussion of repulsive surfaces splits into two categories:

(i)   where the minimum energy path corresponding to the "traditional reaction coordinate" is taken to be the most probable route (Figure 6.4c). This is relevant to situations where the attacking atom is light.
(ii) where a "cutting the corner" trajectory is the most probable route (Figure 6.4d). This is relevant to situations where the attacking atom is heavy.

The actual potential energy surfaces are the same for both the $L + HH$ and $H + LH$ cases with both being repulsive with a late energy barrier. The equilibrium internuclear distances $(r_{AB})_{eq}$ and $(r_{BC})_{eq}$ are shown in Figures 6.10 and 6.11.

For the minimum energy path, r_{AB} for the critical configuration at the col representing the activated complex is close to $(r_{AB})_{eq}$, but r_{BC} is much larger than $(r_{BC})_{eq}$.

The "cutting the corner" trajectory is displaced from the minimum energy path, so that the activated complex configuration is displaced upwards and r_{AB} at the highest point on the trajectory is now larger than $(r_{AB})_{eq}$.

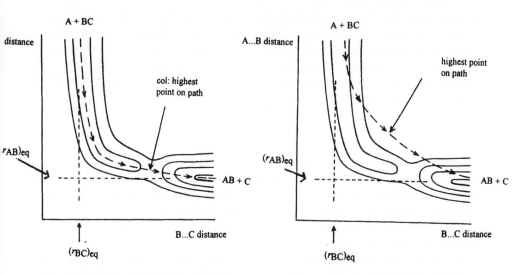

Figure 6.10 A potential energy contour diagram showing positions of $(r_{AB})_{eq}$ and $(r_{BC})_{eq}$ for the minimum energy path.

Figure 6.11 A potential energy contour diagram showing positions of $(r_{AB})_{eq}$ and $(r_{BC})_{eq}$ for the "cutting the corner" path.

When actual trajectory calculations are considered, it is only when the attacking atom is light that the minimum energy path is followed

$$L + HH \quad \rightarrow \quad LH + H$$
$$L + HL \quad \rightarrow \quad LH + L$$
$$L + LH \quad \rightarrow \quad LL + H$$

When the attacking atom is heavy, trajectory calculations indicate that the "cutting the corner" trajectory now becomes the most probable route

$$H + HL \quad \rightarrow \quad HH + L$$
$$H + LH \quad \rightarrow \quad HL + H$$
$$H + LL \quad \rightarrow \quad HL + L$$

The cases

$$L + LL \quad \rightarrow \quad LL + L$$
$$H + HH \quad \rightarrow \quad HH + H$$

are not dealt with here. In these reactions all atoms are of the same mass type, and their behaviour is similar to $H + HL$ but is less extreme (Section 6.14).

The repulsive surface for the cases of light atom attacking and heavy atom attacking will be discussed separately.

6.12 General features of repulsive potential energy surfaces where the attacking atom is light

On repulsive surfaces when the attacking atom is light the trajectory across the surface is the minimum energy path, Figure 6.10, that is the "traditional" reaction coordinate, the reaction unit A...B...C does not reach the barrier or col until the A...B distance is close to its equilibrium internuclear distance, while the BC bond has extended to a considerable extent. As a result the barrier is placed in the exit valley. For the minimum energy path, the AB bond is virtually formed before B and C have separated much, so that energy release occurs while there is increasing separation of products. The col, therefore, corresponds to a configuration

$$A\text{----}B\text{----------}C \qquad r_{BC} \gg r_{AB}$$

In the exit valley the reaction energy is released as the molecule AB and atom C separate. Because the AB distance is virtually the equilibrium internuclear distance, AB moves away from C as a single entity, and as the repulsion energy from BC is released to this specific geometry there is likely to be only minimal recoil of B into A and the vibrational content of the products will be minimal. There will, therefore, be a high proportion of translational energy in the products.

And so on a repulsive surface with movement along the minimum energy path, these simple considerations lead to the prediction that translational energy is predominant in products.

Repulsive surfaces with late barriers are enhanced by high vibrational energy in the reactants with translational energy playing a minor role.

Translational energy is less important for surmounting the repulsive late barrier because high translational energies would cause the "reaction entity" (A...B...C) to hit the side walls of the entrance valley and be reflected back down the entrance valley (Figure 6.10).

Trajectory calculations predict that there is vibrational enhancement or promotion of reaction for the repulsive surface though at very high vibrational energies the excess vibrational energy above the vibrational threshold value will appear as vibrational energy in the products.

These predictions derive from theoretical potential energy surfaces and from computed trajectories across them. Yet again, molecular beam experiments can test these predictions and the following predictions will hold, *but only for reactions involving a light atom as reactant*.

For the repulsive surface with a light atom attacking, theory and experiment show that the following generalisations hold, *but only for the specific cases of a light atom as reactant*.

1. Vibrational energy enhances reaction.
2. Translational energy is dominant in the products.
3. Selective enhancement by vibration is best when there is a *curved* route up the entrance valley to the critical configuration of the activated complex at the col. Vibration can enable the "reaction unit" (A...B...C) to get round the bend easily. High translational energies in the reactants would cause the unit (A...B...C) to hit into the side wall at the end of the valley, and be reflected back along the reactant valley.

4. Product energy is predominantly translational when there is a *straight* route from the *entrance to the exit valley* up to the critical configuration at the col.
5. This can again be put in terms of a momentum argument. When the vibrational phase is such as to permit the unit (A...B...C) to round the bend, transverse motion, that is vibration, in the entrance valley is converted to parallel motion, that is translation which will take it down the exit valley.

Molecular beam experiments again complement these conclusions. Repulsive surfaces are associated with the backward scattering of a rebound mechanism in which A collides with BC in a head on collision and AB rebounds backwards. Head on collisions are associated with small cross sections and small impact parameters where AB has formed as C splits off and separates from B under short range repulsion.

6.13 General features of repulsive potential energy surfaces for exothermic reactions where the attacking atom is heavy

The behaviour for reactions where the attacking atom is heavy, such as $H + HL$, or $H + LL$, is very different from that predicted above, for reactions where the attacking atom is light. Selective enhancement by vibration is both predicted and observed for reactions with a light atom L as a reactant, but where the attacking atom is heavy both vibration and translation are effective in promoting reaction, though vibrational energy is the more important. Similarly, there is a difference in the energy disposal in the products. For example, in the $L + HH$ reaction, translational energy is dominant in the products with very little vibrational energy appearing in BC, but for the cases involving a heavy atom attacking, such as $H + HL$, translation is still dominant, but a *much larger* fraction of the energy is vibrational compared to the situations when light atoms attack.

This is interpreted in terms of the reaction unit "cutting the corner", Figure 6.11 which is an alternative way of saying that the most probable route is now *no longer* the minimum energy path.

This means that when B and C start to separate and move apart they do so before the AB bond is at its equilibrium internuclear distance, that is bond A-B is forming while B-C is breaking

$$\rightarrow \qquad \leftarrow \qquad\qquad \rightarrow$$
$$\text{A-------- B------------------- C}$$

This means that AB is not moving apart from C as a single entity, and to some extent B is being propelled towards A while the A-B bond is forming, and so there will be some vibrational release instead of the very dominant translational energy release found in the $L + HH$ case, and this is the case *despite* the barrier being located in the exit valley. Both translation and vibration will appear in the products, with the most vibration appearing as the "cutting the corner" trajectory is progressively moved vertically upwards.

These trajectory conclusions are confirmed by chemiluminescence and molecular beam experiments for some reactions with highly repulsive surfaces, but which nonetheless give both vibration and translation in products.

6.14 Other mass type reactions

The two categories of reaction which have been discussed so far are those for

$$L + HH \quad \rightarrow \quad LH + H$$
$$H + HL \quad \rightarrow \quad HH + L$$

It can be shown that the category $L + HL$ and $L + LH$ are similar to $L + HH$ but are less extreme with the ordering being $L + HH$, $L + HL$, $L + LH$. Other combinations are fairly similar to $H + HL$, but are less extreme.

The cases of $L + LL$ and $H + HH$ are similar because the mass ratios for each reaction are very similar, and their behaviour is similar to $H + HL$ but is less extreme.

6.15 Less extreme barriers

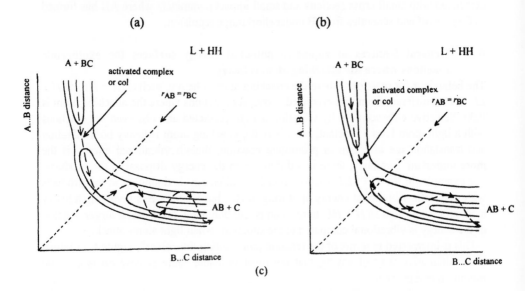

(a)   (b)

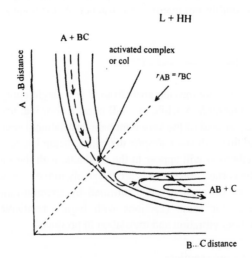

(c)

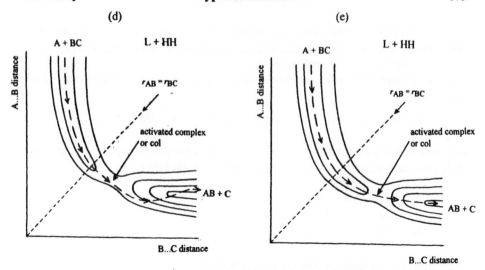

Figure 6.12 Potential energy contour diagrams showing a series of graded surfaces and barriers

(a) an attractive surface, (b), (c) and (d) intermediate surfaces progressing from attractive to repulsive surfaces, (e) a repulsive surface.

Early early barriers, and late late barriers are the extremes of the behaviour described above. The highest percentage of vibrational energy in the products is associated with a really early barrier, while the highest percentage of translational energy in the products is found with a really late barrier. There are also intermediate types of surfaces, and in the progression from early to late barriers a decreasing fraction of vibrational energy, and an increasing fraction of translational energy in the products is predicted from the trajectory calculations. The diagram shows a series of graded surfaces, (a) is an attractive surface with a lot of vibrational energy in the products, (e) is a repulsive surface with a lot of translational energy in the products, and (b), (c) and (d) are intermediate surfaces progressing from attractive to repulsive, with decreasing amounts of vibration in the product.

These diagrams are given for minimum energy paths for cases where a light atom L is the attacking atom. They have to be modified to account for the fact that on repulsive surfaces where the attacking atom, H, is heavy the most probable route is now not the minimum energy path but a "cutting the corner" trajectory.

6.16 Typical reactions for which a correlation of potential energy surfaces and trajectory calculations has been made with molecular beam experiments

A large number of the reactions which have been studied by molecular beams are of the type

$$H + HL \rightarrow HH + L$$

Reactions which have attractive surfaces include alkali metals with halogens and other simple molecules, for example

$$Na \cdot + Cl_2 \rightarrow NaCl + Cl \cdot$$
$$Cs \cdot + I_2 \rightarrow CsI + I \cdot$$
$$Na \cdot + PCl_3 \rightarrow NaCl + P \cdot Cl_2^-$$

These are stripping reactions with large impact parameters and cross sections, and they give forward scattering. The molecular beam experiments confirm the validity of the trajectory predictions for an attractive surface.

When repulsive surfaces are considered, again most of the reactions studied are of the type

$$H + HL \quad \rightarrow \quad HH + L \quad \text{and} \quad H + LH \quad \rightarrow \quad HL + H$$

Reactions of this type occurring on repulsive surfaces between halogen atoms and HCl or HF have been exceptionally well studied, with examples being

$$\begin{aligned}
Br\cdot + HCl \quad &\rightarrow \quad HBr + Cl\cdot \\
Cl\cdot + HF \quad &\rightarrow \quad HCl + F\cdot \\
Br\cdot + HF \quad &\rightarrow \quad HBr + F\cdot
\end{aligned}$$

along with the even more extensively studied reaction

$$K\cdot + CH_3I \quad \rightarrow \quad KI + CH_3\cdot$$

These involve head on collisions with backward scattering occurring at small impact parameters, and having small cross sections. Extensive molecular beam studies confirm the prediction of trajectory calculations over repulsive surfaces for the *H/HL* mass type reactions.

The reaction

$$F\cdot + H_2 \quad \rightarrow \quad HF + H\cdot$$

has been well studied, and an accurate quantum potential energy surface has been calculated. This reaction is of mass type

$$H + LL \quad \rightarrow \quad HL + L$$

and shows some interesting behaviour. The surface is highly repulsive and gives both vibrational and translational energy release, with the translational energy release being dominant as predicted in Section 6.13. At low margins above the threshold energy corresponding to $v = 1$ and $v = 2$, there is the backward scattering expected for repulsive surfaces, and found in actual molecular beam experiments. However, when the energy is even higher, $v = 3$, scattering is now forward, and this is not expected. Resonance effects are believed to cause the change.

There is normally some backward scattering for all types of barrier, and reactions in the *L + HH* and *L + HL* categories give anomalously small amounts of forward scattering on attractive surfaces.

6.17 Endothermic reactions
An endothermic reaction is the exact reverse of the corresponding exothermic reaction, Figures 6.3 and 6.9. An *attractive* surface for the *exo*thermic reaction becomes a *repulsive* surface for the *endo*thermic reaction, while the *repulsive* surface for the *exo*thermic reaction becomes an *attractive* surface for the *endo*thermic reaction.

Likewise, if an exothermic reaction is enhanced by translational energy and has predominantly vibrational energy in the products, then the reverse endothermic reaction will be promoted by vibrational energy, and will have translational energy dominant in the products.

The discussion will be limited to the two mass types:

1) the endothermic reaction $H + HL$ which is the reverse of the exothermic reaction $L + HH$

2) the endothermic reaction $L + HH$ which is the reverse of the exothermic reaction $H + HL$

I *An exothermic reaction on a repulsive surface with a late type II barrier (Figure 6.13)*	I *An endothermic reaction on an attractive surface with an early Type I barrier (Figure 6.14)*
(i) Mass type: $L + HH$	(i) Mass type: $H + HL$

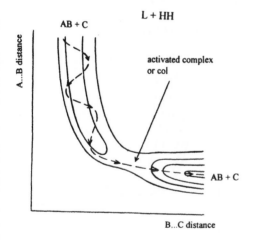

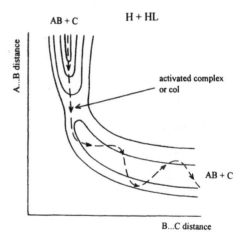

Figure 6.13 An exothermic reaction on a repulsive surface with a late barrier. Mass type: $L + HH$	**Figure 6.14** An endothermic reaction on an attractive surface with an early barrier. Mass type: $H + HL$

(ii) This repulsive surface has a curved run up to the barrier, and so vibrational energy is more effective in getting the "reaction unit" (A...B...C) round the corner. Hence vibrational energy enhances reaction.

(ii) The exothermic reaction requires vibrational energy for promotion of reaction. Hence this endothermic reaction produces predominantly vibrational energy in the products, in keeping with a curved run up the exit valley to the col.

(iii) This repulsive surface produces predominantly translation in the products in keeping with a straight run up the exit valley.

(iii) Since the exothermic reaction gives translational energy in the products, then this endothermic reaction must require translational energy for promotion of reaction, in keeping with a straight run up the entrance valley.

(iv) A late barrier with $r_{BC} > r_{AB}$ means that the bond A-B is formed before the repulsion energy of BC is released giving translational energy predominant in products.

(iv) Having an early barrier with $r_{AB} > r_{BC}$ means that the repulsion energy of BC is released before the bond A-B is formed, causing B to recoil into A giving vibration predominant in the product AB.

II An exothermic reaction on an attractive surface with an early Type I barrier (Figure 6.15)

II An endothermic reaction on a repulsive surface with a late Type II barrier (Figure 6.16)

(i) Mass type: $L + HH$

(i) Mass type: $H + HL$

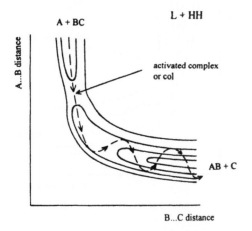

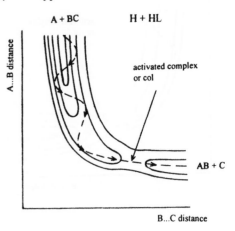

Figure 6.15 An exothermic reaction on an attractive surface with an early barrier. Mass type: $L + HH$

Figure 6.16 An endothermic reaction on a repulsive surface with a late barrier. Mass type: $H + HL$

(ii) This attractive surface has a straight run up to the barrier, and so translational energy promotes reaction

(ii) The exothermic reaction requires translational energy for promotion of reaction. Hence this endothermic reaction produces predominantly translational energy in the products, in keeping with a straight run up the exit valley.

(iii) This attractive surface produces predominantly vibration in products, in keeping with a curved run up the exit valley.

(iii) Since the exothermic reaction gives vibrational energy in the products, then this endothermic reaction must require vibrational energy for promotion of reaction, in keeping with a curved run up the entrance valley.

(iv) Having an early barrier with $r_{AB} > r_{BC}$ means that the repulsion energy of BC is released before the bond A-B is formed, causing B to recoil into A giving vibration predominant in the product AB.

(iv) A late barrier with $r_{BC} > r_{AB}$ means that the bond A-B is formed before the repulsion energy of BC is released giving translational energy predominant in products.

III An exothermic reaction on an attractive surface with an early Type I barrier (Figure 6.17)

III An endothermic reaction on a repulsive surface with a late Type II barrier (Figure 6.18)

(i) Mass type: $H + HL$

(i) Mass type: $L + HH$

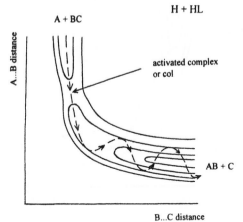

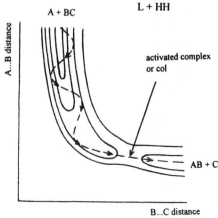

Figure 6.17 An exothermic reaction on an attractive surface with an early barrier. Mass type: $H + HL$

Figure 6.18 An endothermic reaction on a repulsive surface with a late barrier. Mass type: $L + HH$

(ii) This attractive surface has a straight run up the entrance valley, and so translational energy promotes reaction

(ii) The exothermic reaction requires translational energy for promotion of reaction. Hence this endothermic reaction produces translational energy predominant in the products, in keeping with a straight run up the exit valley.

(iii) This attractive surface produces predominantly vibrational energy in the products, in keeping with a curved run up the exit valley.

(iv) Having an early barrier with $r_{AB} > r_{BC}$ means that the repulsion energy of BC is released before the bond A-B has formed, causing B to recoil into A giving vibration in the products.

IV An exothermic reaction on a repulsive surface with a late Type II barrier (Figure 6.19)

(i) Mass type: $H + HL$

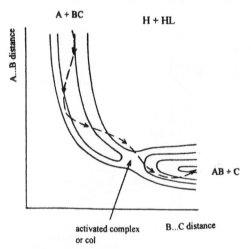

Figure 6.19 An exothermic reaction on a repulsive surface with a late barrier. Mass type: $H + HL$

(ii) This repulsive surface has a curved run up the entrance valley, and so vibrational energy promotes reaction.

(iii) Since the exothermic reaction produces vibrational energy in the products, then this reaction requires vibrational energy to promote reaction, in keeping with the curved run up the entrance valley.

(iv) Having a late barrier with $r_{BC} > r_{AB}$ means that the bond A-B is formed before the repulsion energy is released giving translational energy in the products.

IV An endothermic reaction on an attractive surface with an early Type I barrier (Figure 6.20)

(i) Mass type: $L + HH$

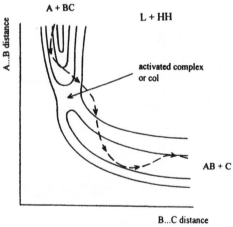

Figure 6.20 An endothermic reaction on an attractive surface with an early barrier. Mass type: $L + HH$

(ii) The exothermic reaction requires vibrational energy for promotion of reaction. Hence this endothermic reaction produces predominantly vibrational energy in the products in keeping with the curved run up the exit valley.

(iii) This repulsive surface has a straight run up the exit valley, and so translational energy could be expected to be predominant in the products. However, the approaching atom A is heavy, and so the most likely trajectory across the repulsive surface is no longer the minimum energy path, but is one which "cuts the corner" and so results in translational energy being still predominant in the products but with a considerable contribution from vibration, *despite* there being a *late* barrier and a straight run up the exit valley.

(iv) Having a late barrier means that $r_{BC} > r_{AB}$, but the fact that the most probable path over the surface is a trajectory which cuts the corner means that the standard arguments as illustrated in cases I, II and III will not apply. The trajectory used lies above the minimum energy pathway. This means that AB is not moving apart from C as a single entity, and to some extent B is being propelled towards A while the A-B bond is still forming, and so there will be some vibrational energy release *despite* the barrier being located in the exit valley (Section 6.14).

(iii) The exothermic reaction proceeds by way of a trajectory which "cuts the corner" and so produces both vibrational and translational energy in the products. Hence the endothermic reaction must also proceed via a trajectory which "cuts the corner" and so both vibrational and translational energy must promote reaction, even though translation will be the major component.

(iv) Having an early barrier means that $r_{AB} > r_{BC}$. But because the trajectory is one which "cuts the corner" this again means that the standard conclusions do not follow. Both translation and vibration appear.

6.18 Thermoneutral reactions

The situation is similar with thermoneutral reactions. If there are late barriers, vibrational energy is the most important for promotion of reaction, while early barriers require translation. In both situations there is again interchange of energy, where

1. vibrational energy of reactants is transferred into translational energy of products,
2. translational energy of reactants is transferred into vibrational energy of products,
3. vibrational and translational energy in excess of the respective threshold values remain as vibrational and translational energy respectively, and appear as such in the products.

However, the same note of caution must again be sounded. Careful assessment of the energy requirements for the endothermic and thermoneutral, $L + HH \rightarrow LH + H$ type reactions must be made in the light of the "cutting the corner" trajectories found in the reverse exothermic $H + HL \rightarrow HH + L$ type reaction.

6.19 Exo and endothermic reactions, attractive and repulsive surfaces

A rough and ready rule suggests that endothermic reactions tend to occur on repulsive surfaces, highly exothermic reactions on attractive surfaces, and exothermic reactions on both.

The potential energy profiles can help to explain this.

Figure 6.21a shows how configurations of a reaction unit (A...B...C) with $r_{AB} = r_{BC}$ lie along the line XY. A symmetrical barrier will have the critical configuration lying somewhere along XY. An early barrier has a critical configuration which lies above the line XY, and a late barrier has one lying below XY. On the potential energy profile Figure 6.21b the line XY can be drawn in, and the early barrier will lie to the left of XY, and the late to the right of XY.

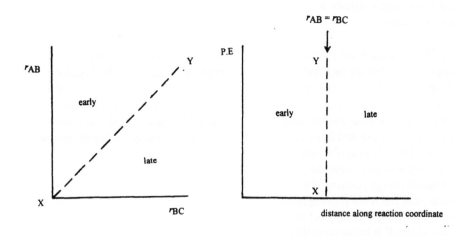

Figure 6.21a Early and late barriers in tems of configuration of the "reaction unit",   **Figure 6.21b Placing of barriers on the potential energy profiles.**

In highly endothermic reactions and highly exothermic reactions there is a long smooth climb up to the critical configuration on one side, and a short sharp climb on the other, Figures 6.22a and b. In slightly endothermic and slightly exothermic reactions the climbs are comparable on each side, Figures 6.23a and b.

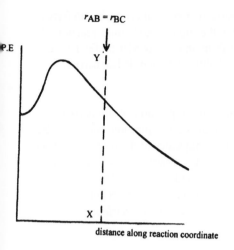

Figure 6.22a Placing of the barrier for a highly exothermic reaction

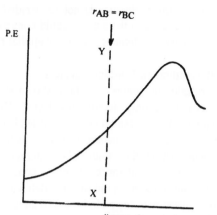

Figure 6.22b Placing of the barrier for a highly endothermic reaction.

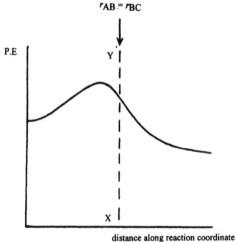

Figure 6.23a Placing of the barrier for a slightly exothermic reaction

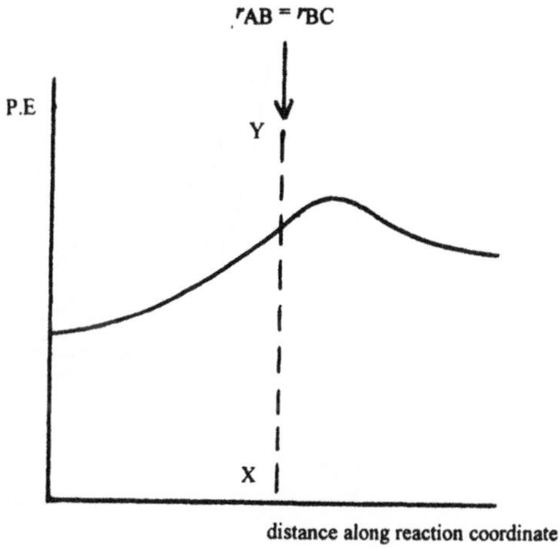

Figure 6.23b Placing of the barrier for a slightly endothermic reaction.

The line XY representing $r_{AB} = r_{BC}$ is put on to these profiles somewhere along the long smooth incline. This is unambiguous for the highly exo or highly endothermic reactions, but for the slightly exo or slightly endothermic reactions the line could easily be placed to either side. As indicated previously, a line to the right of the maximum defines an early barrier, one to the left defines a late barrier.

The conclusions are:

1. a highly exothermic reaction favours an early barrier,
2. a highly endothermic reaction favours a late barrier

3. but for the slightly exothermic reaction an early barrier is likely to be preferred,
    though the preference is not so decided as for the highly exothermic reaction. A
    small change in the surface could convert this from an early barrier preference to a
    late barrier preference. The same applies to the slightly endothermic late barrier.

6.20 Gradual and sudden barriers

The reaction dynamics have been shown to depend on the position of the barrier, but
this is not the only determining factor. The curvature of the reaction path and the
steepness of the potential energy profile also affect the trajectory of the reaction unit on
the potential energy surface. Calculations have shown that trajectories over the surface
are dependent on the amount of curvature of the path around the col, and that when this
is taken into consideration there is a considerable improvement in the accuracy of the
predictions made from the surface. Behaviour on gradual and sudden surfaces is related
to the curvature of the path, Figures 6.24 and 6.25.

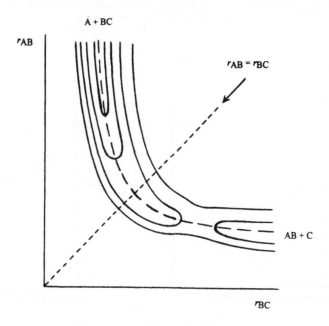

Figure 6.24 A gradual potential energy contour diagram.

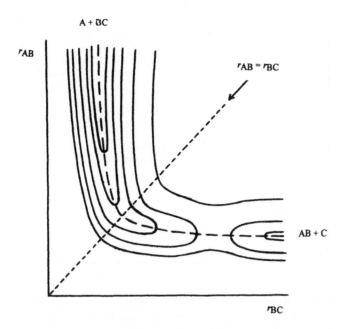

A + BC

r_{AB}

$r_{AB} = r_{BC}$

AB + C

r_{BC}

Figure 6.25 A sudden potential energy contour diagram.

In the gradual potential energy profile, Figure 6.24 the curvatures are such that there is a much clearer straight run up to the barrier, with only a slight bend at the end of the valley. This has to be compared with the more curved sudden potential energy profile, Figure 6.25. Vibration is more effective for movement up a valley with a bend at the end, and so the sudden profile is much more easily surmounted by vibrational energy.

In a gradual surface, or profile, the potential energy starts to rise at a very early stage of the reaction, and so the base of the barrier is also encountered at an early stage. This early part of increasing potential energy is best surmounted by translational energy, but both translation and vibration are effective in enabling the barrier to be crossed, Figure 6.26.

In a sudden surface, or profile, the potential energy hardly changes during the early stages of reaction, and the barrier is encountered much further along the reaction path, but where the potential energy has still only changed slightly. There is then an abrupt rapid change in potential energy, and in this situation vibrational energy is more effective for crossing the barrier, Figure 6.27.

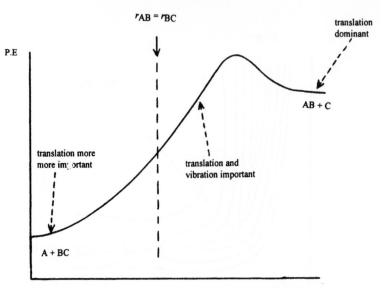

Figure 6.26 A gradual potential energy profile showing the importance of translational and vibrational energy at various stages along the reaction coordinate for a highly endothermic reaction with a late barrier.

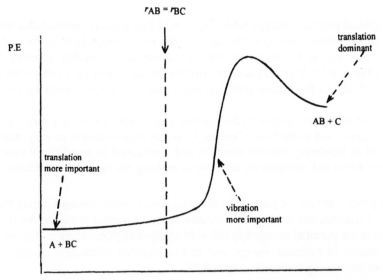

Figure 6.27 A sudden potential energy profile showing the relative importance of translational and vibrational energy at various stages along the reaction coordinate for a highly endothermic reaction with a late barrier.

6.21 Potential energy wells

Sometimes a basin or well is found somewhere along the reaction coordinate, Figure 6.7. If the well is in the entrance valley the "reaction unit" needs vibrational energy, but if it is in the exit valley translational energy is more effective. Although wells generally mean collision complexes, this need not be the case.

7

Simple and Modified Collision Theory

In simple collision theory reaction occurs when two molecules which have energy greater than a certain critical value collide. This requires calculation of two quantities:
(i) the *total rate* of collision of the reactant molecules,
(ii) the *fraction* of molecules which have at least a certain critical value of the energy.

The resultant equation can be compared with the empirical Arrhenius equation

$$k = A\exp\left(-\frac{E}{RT}\right).$$

(7.1)

Simple collision theory assumes that the colliding molecules are hard spheres, and ignores the molecular structure and details of internal motion such as vibration and rotation. Likewise it does not consider the products of reaction nor their internal structure.

Modern collision theory which is based on molecular beams and laser induced fluorescence studies, looks at a collision in much more detail. Now the internal states of the colliding molecules, the internal states of the products of a collision and the distribution of energy among the product molecules become of fundamental importance.

7.1 Simple collision theory
Reaction can be between like molecules

$$A + A \rightarrow products$$

and the total rate of collision is modified to account for the fact that only those collisions which have the critical energy or greater can lead to reaction.

When reaction is between two unlike molecules

$$A + B \rightarrow products$$

only collisions between A and B have any chance of leading to reaction, so that the appropriate total rate of collision is for collisions between A and B only. Collisions between A and A, or B and B do not count as they can never lead to reaction. Again the appropriate collision rate has to be modified to account for the fact that not all collisions will cause reaction.

7.1.1 Definition of a collision

Molecule A approaches molecule B with relative velocity, v_{AB}, conveniently represented by considering B to be stationary and A to approach B with velocity v_{AB}.

$$v_{AB}$$
$$\text{moving A} \quad \rightarrow \quad \text{stationary B}$$

A collision will only occur if the line representing the approach of A towards B is at a distance less than $r_A + r_B$ from the centre of B, r_A and r_B being the radii of molecules A and B respectively. A and B will not collide if the centre of B lies at a distance greater than $r_A + r_B$ from the line representing the motion of A towards B, Figure 7.1.

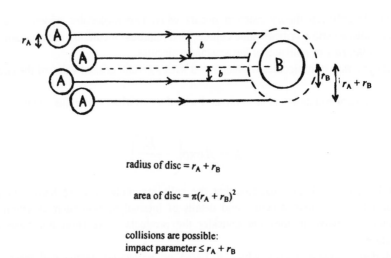

radius of disc $= r_A + r_B$

area of disc $= \pi(r_A + r_B)^2$

collisions are possible:
impact parameter $\leq r_A + r_B$

Figure 7.1 Criteria for collision: a possible collision

The distance b shown in Figure 7.1 is called the impact parameter. A head on collision occurs when the impact parameter, b, is zero, and collisions will occur if the impact parameter is equal to or less than $r_A + r_B$. There will be no collision if the impact parameter is greater than $r_A + r_B$.

This criterion of a collision can be given in terms of an effective collision diameter d_{AB} where $d_{AB} = r_A + r_B$. Collisions will not occur if the impact parameter is greater than d_{AB}, but will occur if the impact parameter is equal to or less than d_{AB}.

A third criterion is in terms of a cross sectional area, σ, for collision. If a B molecule is placed at the centre of a disc of cross section σ, then all A molecules whose centres hit anywhere on the disc must collide with the B molecule at the centre of the disc. Any A which ends up by hitting outside this disc will not collide with B. The area of this disc is defined in terms of the radii of the colliding molecule and is $\pi (r_A + r_B)^2$, and this is then the cross sectional area σ, Figure 7.2.

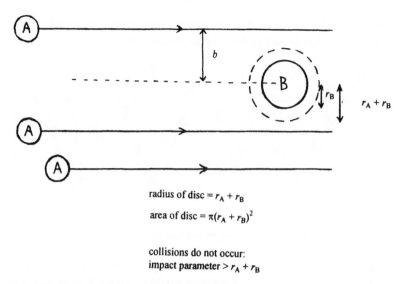

radius of disc $= r_A + r_B$

area of disc $= \pi(r_A + r_B)^2$

collisions do not occur:
impact parameter $> r_A + r_B$

Figure 7.2 Criteria for collision: no collision possible

Simple collision theory formulations generally use the sum of the radii or an effective collision diameter, cross sections tend to be reserved for molecular beam studies and modified collision theory.

7.1.2 Formulation of the total collision rate

There are several well defined steps leading to a calculation of the total collision rate.
(a) Encounters with impact parameter $b < r_A + r_B$ count to the collision rate, those where $b > r_A + r_B$ do not. This criterion defines the radius of a cylinder, that is $r_A + r_B$, which encloses all collision partners, so that all molecules within the cylinder will collide with each other, but molecules outside the cylinder will not collide with those inside the cylinder.

This cylinder is build up from a sequence of cross sections each of which has the area of a disc, and so the cross section simply represents a slice through a cylinder. A cylinder of collisions is generated since all A molecules hitting the disc around the B molecule have travelled towards B through a succession of discs of the same radius. The radius of the cylinder is $r_A + r_B$ since this is the radius of the cross section $\pi(r_A + r_B)^2$, Figure 7.3.

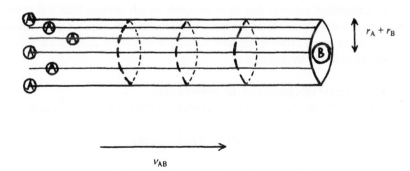

v_{AB}

Figure 7.3 A cylinder defining collisions.

(b) A similar cylinder can be used to describe the total number of collisions occurring between A and B molecules within the cylinder, Figure 7.4.

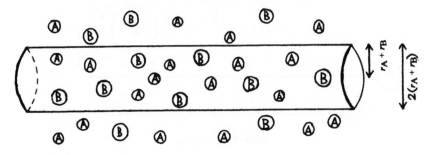

Figure 7.4 Number of collisions occurring in one second.

One of the A molecules in the cylinder can be chosen as a reference molecule. As this reference molecule moves down the cylinder it would hit all the B molecules within the cylinder, but would not collide with any B molecules outside the cylinder. It would also hit A molecules, but these collisions are not pertinent to a calculation of the rate of collision of A with B molecules.

The reference A molecule moves down the cylinder of radius $r_A + r_B$ with relative velocity v_{AB}. In one second A travels a distance v_{AB} along the cylinder, and so a cylinder of length v_{AB} and radius $r_A + r_B$ represents a region which encloses all the B molecules, and incidentally A molecules, which are hit as A moves down the cylinder.

(c) The molecules are considered to be hard spheres for which there will be no interactions at distances greater than $r_A + r_B$. If there were interactions at distances greater than $r_A + r_B$ the molecules would no longer be hard spheres, and encounters at distances greater than $r_A + r_B$ would be collisions. A hard sphere treatment is simplistic and physically naive, and modern studies based on molecular beams tacitly assume that there are interactions at distances greater than $r_A + r_B$.

(d) The cylinder described in step (b) contains all the A and B molecules which are hit by the moving reference molecule A. The number of collisions between A and B per unit time per unit volume can be found by counting up the number of B molecules in this cylinder which has volume $\pi(r_A + r_B)^2 \times v_{AB}$. A similar calculation would give the collision rate for collisions between A and other A molecules, and between B and other B molecules, but as these can never lead to reaction they are irrelevant.

The calculation assumes that an A molecule which collides with a B molecule will move on in a straight line to hit the next B, and the next B. This is a gross simplification, and molecular beam experiments look in detail at how the A molecule is deflected from its original pathway by collision.

7.1.3 The actual calculation of the collision rate

The number of B molecules inside the cylinder of volume $\pi(r_A + r_B)^2 v_{AB}$ is

$$\pi\left(r_A + r_B\right)^2 v_{AB} N_B$$

where N_B is the number of B molecules per unit volume.

In one second the reference A molecule will have traversed the length of the cylinder and will hit all the B molecules in the cylinder, that is the number of collisions between the reference molecule A and all the B molecules is $\pi(r_A + r_B)^2 v_{AB} N_B$.

But any of the A molecules inside the cylinder could be considered to be the reference A molecule. As there are N_A molecules of A per unit volume, then the total number of collisions per unit time per unit volume is

$$\pi(r_A + r_B)^2 v_{AB} N_A N_B. \tag{7.2}$$

The average relative velocity v_{AB} can be calculated from kinetic theory where

$$v_{AB} = \left(\frac{8kT}{\pi\mu}\right)^{\frac{1}{2}} \quad \text{where} \quad \frac{1}{\mu} = \frac{1}{m_A} + \frac{1}{m_B}. \tag{7.3}$$

The total collision rate per unit time per unit volume is therefore

$$\left(\frac{8\pi kT}{\mu}\right)^{\frac{1}{2}} (r_A + r_B)^2 N_A N_B. \tag{7.4}$$

This collision rate is often called "the collision frequency", $\mathcal{Z}_{AB}$ where

$$\mathcal{Z}_{AB} = Z_{AB} N_A N_B \tag{7.5}$$

where Z_{AB} is a constant of proportionality, called the collision number, and

$$Z_{AB} = \left(\frac{8\pi kT}{\mu}\right)^{\frac{1}{2}} (r_A + r_B)^2. \tag{7.6}$$

Expressed in terms of the effective collision diameter $d_{AB} = r_A + r_B$ the collision number becomes

$$\left(\frac{8\pi kT}{\mu}\right)^{\frac{1}{2}} d_{AB}^2 \tag{7.7}$$

and in terms of the cross sectional area $\sigma = \pi(r_A + r_B)^2$ is

$$\left(\frac{8kT}{\pi\mu}\right)^{\frac{1}{2}} \sigma. \tag{7.8}$$

7.1.4 Calculation of the total rate of reaction

The collision rate for A and B molecules calculated above in general is much greater than the observed rate of reaction. This is because not all collisions lead to reaction, only those which have at least a certain critical energy will be successful.

For a bimolecular reaction this critical energy corresponds to the kinetic energy of relative translational motion along the line of centres of the colliding molecules, often loosely described as the "violence" of the collision.

$$A \quad \bullet \; \rightarrow \qquad\qquad \leftarrow \; \bullet \quad B$$

kinetic energy = $\frac{1}{2} m v_A^2$ kinetic energy = $\frac{1}{2} m v_B^2$

and $\frac{1}{2} m v_A^2 + \frac{1}{2} m v_B^2 \quad \gg \quad$ a critical value

This is an energy in 2 squared terms v_A^2 and v_B^2.

The Maxwell-Boltzmann distribution gives "the fraction of molecules which have energy at least ε in two squared terms" as

$$\exp\left(-\frac{\varepsilon}{kT}\right) \quad \text{where} \quad \varepsilon = \frac{1}{2} m v_A^2 + \frac{1}{2} m v_B^2. \tag{7.9}$$

The total rate of reaction is the total rate of collision of A and B molecules, modified by this term to give the total rate of successful collisions as

$$\left(\frac{8\pi kT}{\mu}\right)^{\frac{1}{2}} (r_A + r_B)^2 \, N_A N_B \exp\left(-\frac{\varepsilon}{kT}\right) \tag{7.10}$$

$$= Z_{AB} \exp\left(-\frac{\varepsilon}{kT}\right) \tag{7.11}$$

$$= Z_{AB} N_A N_B \exp\left(-\frac{\varepsilon}{kT}\right). \tag{7.12}$$

The actual rate of reaction = $k_{calc} \, N_A N_B$

$$\therefore \quad k_{calc} = Z_{AB} \exp\left(-\frac{\varepsilon}{kT}\right). \tag{7.13}$$

and this calculated rate constant k_{calc} can then be compared with the experimental k in the Arrhenius equation

$$k_{exp\,tl} = A \exp\left(-\frac{E_A}{RT}\right), \tag{7.14}$$

so that the experimental A factor corresponds to the theoretical collision number Z_{AB}, and the experimental activation energy E_A corresponds to the critical minimum energy ε where $\varepsilon = \frac{1}{2} m v_A^2 + \frac{1}{2} m v_B^2$.

Comparison of the calculated Z_{AB} with the experimental A factor shows that for very many reactions Z_{AB} and A have very similar values, but there are some reactions which show differences ranging over several powers of ten. These discrepancies require postulating a "p" factor limiting the theoretical rate of reaction, and whose magnitude can be found from experiment.

$$k = pZ_{AB} \exp\left(-\frac{\varepsilon}{kT}\right)$$

$$= A \exp\left(-\frac{\varepsilon}{kT}\right) \tag{7.15}$$

so that

$$p = \frac{A}{Z_{AB}}. \tag{7.16}$$

The "p" factor has often been interpreted as a preference for a certain direction, or angle of approach, of the two reacting molecules. Molecular beam experiments have shown that some reactions have a quite decided preference for a specific direction and angle of approach, but the need for a "p" factor could also be a reflection of the physically naive model used. This possibility is reinforced by the observation that some "p" factors are greater than unity, and often as large as 10^5 for some reactions in solutions. This is inconsistent with the concept "p" as a factor limiting the rate of successful collisions.

7.1.5 A more realistic criterion for successful collisions

Accumulation of energy in two squared terms is the simplest way in which activation energy can be acquired. But molecules do have internal structure, and the vibrational and rotational states could influence the ease of reaction. Molecular beam studies show that accumulation of vibrational and/or rotational energy may be important. If so, activation energy would be accumulated in $2s$ squared terms, each vibration contributing two squared terms, and each rotation one squared term. The Maxwell-Boltzmann distribution shows that the fraction of molecules with energy at least ε in $2s$ squared terms is approximately

$$\frac{1}{(s-1)!}\left(\frac{\varepsilon}{kT}\right)^{s-1} \exp\left(-\frac{\varepsilon}{kT}\right) \tag{7.17}$$

giving the rate of reaction

$$= \left(\frac{8\pi kT}{\mu}\right)^{1/2} (r_A + r_B)^2 N_A N_B \frac{1}{(s-1)!}\left(\frac{\varepsilon}{kT}\right)^{s-1} \exp\left(-\frac{\varepsilon}{kT}\right) \tag{7.18}$$

$$k = Z_{AB} \frac{1}{(s-1)!}\left(\frac{\varepsilon}{kT}\right)^{s-1} \exp\left(-\frac{\varepsilon}{kT}\right). \tag{7.19}$$

This form of simple collision theory is very rarely used, the simple exponential term being used almost exclusively. A treatment in $2s$ squared terms is rather inconsistent with an assumption of colliding hard spheres with no internal structure. Molecular beams and modern "physical" kinetics remove this inconsistency, because there the internal states are assumed to influence both the collision and reaction rates.

7.1.6 Reaction between like molecules
Reaction is now

$$A + A \quad \rightarrow \quad \text{products}$$

and the derivation is similar to that given previously for reaction of A and B.

The rate of	=	the total number of collisions per unit time per
collision		unit volume

$$= \frac{1}{2}\pi(r_A + r_A)^2 v_{AA} N_A^2$$

$$= 2\pi r_A^2 v_{AA} N_A^2. \tag{7.20}$$

The factor ½ comes in to ensure that each A molecule is not counted twice - each A can be considered as both "hitting" another A molecule, and being "hit" by another A.

Substitution for v_{AA} gives total collision rate, Z_{AA}

$$Z_{AA} = \left(\frac{32\pi kT}{\mu}\right)^{\frac{1}{2}} r_A^2 N_A^2 \tag{7.21}$$

$$= Z_{AA} N_A^2. \tag{7.22}$$

Taking into account that it is only collisions between A molecules having energy at least ε in two squared terms which can react, gives

$$\text{rate of reaction} = Z_{AA} N_A^2 \exp\left(-\frac{\varepsilon}{kT}\right) \tag{7.23}$$

$$\text{and } k = Z_{AA} \exp\left(-\frac{\varepsilon}{kT}\right). \tag{7.24}$$

The corresponding equation for energy at least ε to be accumulated in $2s$ squared terms is

$$k = Z_{AA} \frac{1}{(s-1)!}\left(\frac{\varepsilon}{kT}\right)^{s-1} \exp\left(-\frac{\varepsilon}{kT}\right). \tag{7.25}$$

7.2 Modified collision theory
Simple collision theory has now been replaced by a much more physically realistic approach. One of the most important developments of modern physical chemistry has been the coupling of molecular beam experiments, which get very close to observing what happens in a single collision, with laser techniques which can place molecules in specific quantum states, can identify specific quantum states and can measure the populations of specific quantum states. These studies have provided the main impetus to

the formulation of a modified collision theory. These experiments define collisions and collision processes explicitly, and define in detail the microscopic processes of energy transfer and reaction which can occur on collision. In addition, these techniques give the best data on intermolecular potential energies, and are now used frequently in the construction of potential energy surfaces.

7.3.1 Scattering in a molecular beam experiment
In a molecular beam experiment two low pressure beams, one consisting of A molecules, the other of B molecules, are shot at 90° to each other. Several things can happen at the point of intersection of the beams

1. A molecules continue on undeflected along the original direction of the beam. They are said to be unscattered.
2. B molecules can also be undeflected and unscattered.
3. A and B molecules can interact with each other, and as a result both A molecules and B molecules are deflected from their original trajectories, and both are scattered.

Molecular beam experiments detect scattered and unscattered A and B molecules, and it is inferred that any A or B molecules which have been deflected and scattered must have suffered a collision.

The two beams are both at very low pressures, 10^{-6} or 10^{-7} mmHg is typical. This ensures that there are no collisions between A molecules in the beam of A, and no collisions between B molecules in the beam of B before they intersect. This is essential because in modern experiments each beam is in a selected quantum state, that is it has specific translational, rotational and vibrational energies, and must remain in this selected quantum state until intersection with the other beam. Collisions within either beam A or beam B must not occur, since collisions would cause energy transfer and consequent alteration of the selected quantum states. Likewise there must be no collisions after intersection, otherwise the quantum states of the products would not be preserved.

7.3.2 Detection of scattering
A detector moves around the point of intersection of the beams and defines a sphere around the point of intersection. This can be thought of as a hollow ball with two beams intersecting at 90° to each other exactly at the centre of the ball, Figure 7.5.

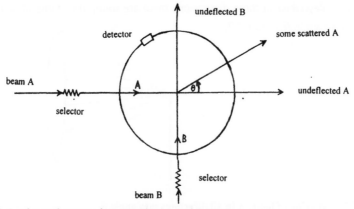

Figure 7.5 A schematic scattering experiment and detector.

The detector moves across the "inside surface of this sphere" and registers the appearance and amount of scattered A and B at positions all over this sphere. Undeflected A and B molecules will be found at a position dictated by extension of their original trajectories. The detector must be sensitive to low numbers of scattered molecules, and also be able to determine the quantum states of the scattered molecules. Modern detectors use mass spectrometry and laser induced fluorescence for detection and determination of quantum states. The total scattering and the scattering into a specific angle can be found, the latter being the most useful, but also the most demanding on the sensitivity of the detector since there are only small numbers of scattered molecules landing up into each specific angle.

The time of flight mass spectrometer measures the velocity distribution of the scattered molecules. A pulse of scattered molecules is admitted, and its intensity measured as a function of flight time down the mass spectrometer. The time to go a given distance gives the velocity, and the intensity at points along the tube at given times gives the number of molecules with a given velocity. A complete analysis gives the velocity distribution, and hence the distribution of translational energy.

7.3.3 Scattering into an angle θ

A collides with B and as a result is scattered at an angle θ to the original direction, Figure 7.6.

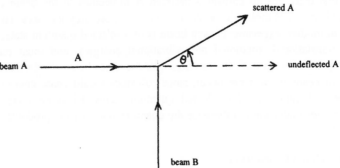

Figure 7.6 Scattering of beam A into angle θ

When this is described in three dimensions there are *many* directions of scattering all with the *same* angle θ, Figure 7.7.

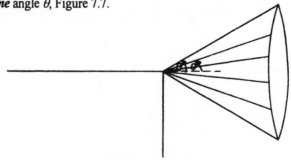

Figure 7.7 Scattering of beam A in all directions into angle θ.

These different directions of scattering into an angle θ define a cone of scattering, and this cone intersects the "inner surface of the sphere" on the rim of a disc lying on this "surface" defined by the moving detector, Figure 7.8.

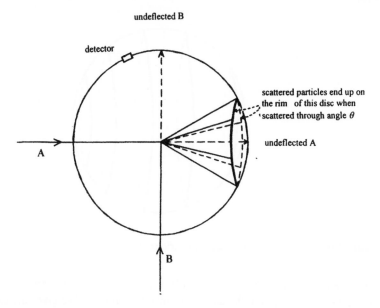

Figure 7.8 Discs of scattering.

But many different angles of scattering are possible, and are observed. The scattered particles, therefore, end up on the rims of a succession of discs each of which corresponds to scattering into a particular angle θ, Figure 7.9.

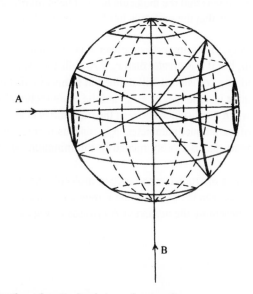

Figure 7.9 Detection of scattering into various angles.

All angles of scattering, 0° to 180°, are possible, and each disc is related to a given angle θ by the geometry of the apparatus. In effect, the sphere can be sliced up into a sequence of infinitesmally narrow discs over the sphere defined by the moving detector, Figure 7.10.

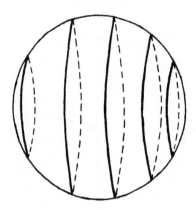

Figure 7.10 Detection of scattering into various angles.

The smallest circumference corresponds to zero or 180° scattering, the largest disc to 90° scattering. Measuring the amount of material on each disc gives the experimenter a value of the amount of material scattered into the corresponding angle θ. The total scattering θ must then be the sum of all scattering into all angles θ.

7.4 Definition of a collision in molecular beam studies

Scattering or deflection is equated to a collision,
(a) no scattering observed means that the molecule has not been involved in a collision,
(b) scattering observed means that the molecule has been involved in a collision.

In simple collision theory, if the impact parameter b is less than or equal to $r_A + r_B$ then a collision will occur, but if b is greater than $r_A + r_B$ collision cannot occur.

In molecular beam studies a collision is still defined in terms of a distance apart of the trajectories, but this distance is no longer $r_A + r_B$, and the cross-sectional area is not $\pi(r_A + r_B)^2$. Collision occurs when the distance between the trajectories is less than or equal to a certain critical value of the impact parameter, b_{max}, and its numerical value is found from scattering experiments and is closely related to the minimum angle of scattering able to be detected.

The impact parameter, b, is of crucial importance in molecular beam experiments, and is defined as the distance between the centre of the B molecule which is considered to be stationary and the line representing the motion of A towards B, Figure 7.11.

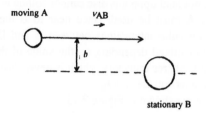

Figure 7.11 Definition of the impact parameter b.

There are a very large number of values of the impact parameter lying in a range of values from $b = 0$ to $b = \infty$, Figure 7.12a.

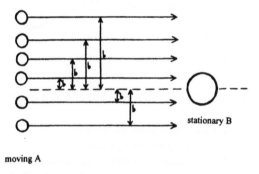

Figure 7.12a Diagram showing various impact parameters.

Within this large range of possible impact parameters there will be a critical range of values which result in a collision. If $b = 0$ there will be a head on collision and if b is greater than the critical upper value to b, b_{max}, collision cannot occur.

This value b_{max} is therefore the radius of a disc drawn around B which represents encounters which lead to collision, and the area of this disc πb^2_{max} is then the cross sectional area for collision, Figure 7.12b.

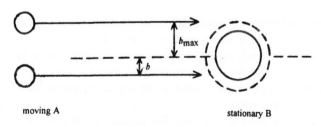

collisions are possible between $b = 0$ and $b = b_{max}$

cross sectional area $\sigma = \pi b^2_{max}$

Figure 7.12b Diagram showing various impact parameters, up to b_{max}.

Any A whose initial trajectory lands up on the disc will collide with B, and any initial trajectory for A which does not land upon this disc cannot collide with B.

The *initial* trajectory of A must be used, since near to B the trajectory becomes distorted because of intermolecular interactions between A and B. These interactions can be repulsive, attractive or mixed depending on the value of the impact parameter. This is in contrast to the hard spheres trajectory which is never distorted before collision because there are no interactions outside $r_A + r_B$.

These trajectory distortions are shown in Figure 7.13.

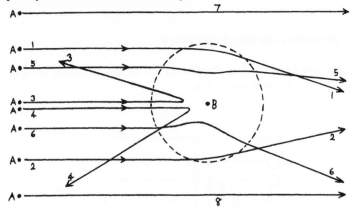

Figure 7.13 Diagram illustrating various trajectories of A as it approaches B.

(a) Trajectories 1 and 2 are dominated by attractions.
(b) Trajectories 3 and 4 are dominated by repulsions.
(c) Trajectories 5 and 6 result from a combination of attractions and repulsions.
(d) Trajectories 7 and 8 do not land up on the disc and will, therefore, not result in collision.

A collision occurs if the impact parameter lies between $b = 0$ and $b = b_{max}$. One of the biggest problems in molecular beam studies is to assign a numerical value to b_{max}. There are no straightforward theoretical calculations which can assign an unambiguous value to b_{max}, which therefore has to be determined experimentally. The magnitude of the experimental value of b_{max} depends crucially on the smallest angle of deflection able to be observed by the particular apparatus, and because of this the value of b_{max} cannot be unique and precise.

7.4.1 The relation between θ and b, estimation of the value of b_{max}, and a definition of a collision in terms of the angle of scattering

There is no way of identifying individual impact parameters prior to an encounter, which is unfortunate since the intermolecular potential energy for interactions between A and B can be deduced for each value of the impact parameter leading to collision. Fortunately, it is also fundamental that each impact parameter is associated with a particular value of the angle of scattering. For instance, $b = 0$ corresponds to a head-on collision which results in totally backwards scattering at 180°, whereas if b is large, but less than b_{max}, interactions will be small resulting in a small angle of scattering. Intermediate situations between $b = 0$ and $b = b_{max}$ result in stronger interactions leading to scattering into larger angles between 0° and 180°, Figure 7.14. This holds in general, except at values of the impact parameter at which attractions are starting to dominate over repulsions, when a mixed situation occurs.

Intermediate situations.

Small b means a large scattering angle.

small b

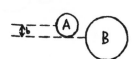

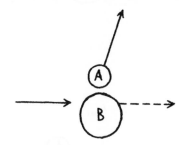

Larger b means a smaller scattering angle.

larger b

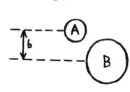

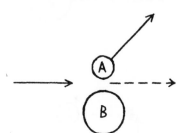

large b

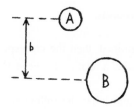

Even larger b means an even smaller scattering angle.

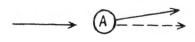

Extreme situations.

$b=0$

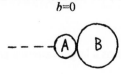

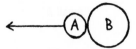

Scattering totally backwards through 180°
- a head-on collision.

$b = b_{max}$

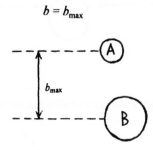

Just made it as a collision;
minimum angle of scattering observed.

$b > b_{max}$

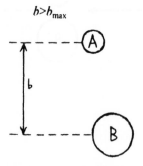

Impact parameter greater than the value
defining a collision, molecule A is undeflected
and there is no collision.

Figure 7.14 Dependence of angle of scattering on impact parameter.

If the angles of scattering, θ, can be precisely determined, then the corresponding values of the impact parameter, b, can be inferred **provided** a relation between θ and b can be derived.

Experimental results show that the graph relating θ and b has the following shape, Figure 7.15.

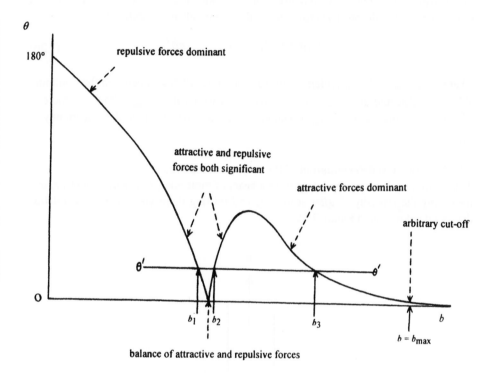

Figure 7.15 A graphical representation of the dependence of θ on b.

Repulsion is dominant at low impact parameters leading to a high value of the scattering angle, while attraction is dominant at large values of the impact parameter leading to smaller angles of scattering. Where attractions and repulsions are balanced, zero scattering results. One problem associated with interpretation of scattering angles is that sometimes a single scattering angle is a result of different impact parameters. For instance scattering into angle θ' results from three different impact parameters b_1, b_2 and b_3.

At very high values of the impact parameter the graph of θ against b approaches zero asymptotically. Zero scattering corresponds to zero deflection which could be taken to represent the boundary between a collision and no collision. But, in fact, zero scattering angle occurs at infinity, and so this cannot be used to distinguish between collisions and no collisions. A compromise is made, and the value of b_{max} which corresponds to the smallest angle at which the detector can detect scattering is taken to represent the division between a collision and no collision. There is, therefore, no definite cut-off in what is taken to be a collision, and so the value of b_{max} and the definition of a collision are fuzzy, and would vary if the minimum angle able to be measured were decreased. However, when this is described in quantum terms the arbitrariness largely disappears.

7.4.2 Collisions and the cross section

In molecular beams a collision occurs when the impact parameter, b, is less than the value, b_{max}, which corresponds to the minimum angle of scattering able to be detected. This corresponds to the radius of a disc over which the interactions of the molecules can be regarded as a collision. The cross sectional area for collision is defined as

$$\text{Area of the disc} = \sigma_{total} = \pi b_{max}^2 \qquad (7.26)$$

where the subscript in σ_{total} refers to all scattering over all angles between the minimum able to be detected and 180°. If σ_{total} can be measured then b_{max} follows. If there is "fuzziness" in the value of b_{max}, (Section 7.4.1), there will be a similar uncertainty in σ_{total}.

7.5 Experimental determination of the total cross section, σ_{total}

This involves measuring the intensity of a beam of molecules, I^o, *before* scattering, and measuring the intensity, I *after* scattering, and relating these via a Beer's law relation described below and in Figure 7.16.

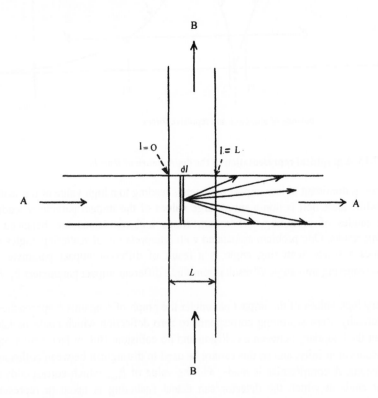

Figure 7.16 Diagram illustrating use of Beer's Law for determining the cross-section for total scattering.

A and B collide in region L. Very low pressures ensure that this is the only collision which either A or B suffers in the incident beam, and likewise no collisions will occur in the scattered beams before they are detected.

As A passes through B there will be collisions and some of A and B will be scattered. The intensity of beam A, I_A, is $N_A v$ where v is the relative velocity with which A approaches B, and N_A is the number of molecules per unit volume of A.

The loss of intensity of beam A due to scattering over the length, dl, depends on

(1) the intensity, $I_A(l)$
(2) the length, dl
(3) the number of molecules per unit volume of B, N_B
(4) the relative velocity, v.

$$-dI_A(l) \propto N_A v N_B dl$$

$$\propto I_A(l) N_B dl$$

$$= \sigma_{(total)}(v) I_A(l) N_B dl \qquad (7.27)$$

where $\sigma_{total}(v)$ is a constant of proportionality, and is the total cross section for the particular relative velocity v.

The net loss of intensity of beam A is the sum of all scattering from $l=0$ to $l=L$. If I_A° and I_A^{f} are the initial and final intensities between $l=0$ and $l=L$, then

$$-\int_{I_A^0}^{I_A^f} \frac{dI_A(l)}{I_A(l)} = \int_0^L \sigma_{total}(v) N_B dl \qquad (7.28)$$

$$\log_e \frac{I_A^0}{I_A^f} = \sigma_{total}(v) N_B L. \qquad (7.29)$$

It is easy to measure I_A° and I_A^{f}, L, N_B and the relative velocity v for the given experiment.

To obtain a value of σ_{total} independent of velocity, $\sigma_{total}(v)$ is summed over the distribution of velocities used in the various experiments

$$\sigma_{total} = \sum_v \sigma_{total}(v) f(v) dv \qquad (7.30)$$

$$= \int_0^\infty \sigma_{total}(v) f(v) dv. \qquad (7.31)$$

When the total scattering is being measured the following relation holds

$$\left\{ \begin{array}{c} \text{amount of} \\ \text{material} \\ \text{scattered} \end{array} \right\} = \left\{ \begin{array}{c} \text{amount of} \\ \text{material in the} \\ \text{original beam} \end{array} \right\} - \left\{ \begin{array}{c} \text{amount of} \\ \text{material which} \\ \text{is not scattered} \end{array} \right\} \qquad (7.32)$$

The final intensity of the undeflected part of the original beam is a quick route to the total scattering over 360°. This is not possible when measuring angular scattering.

The cross section $\sigma_{total}(v)$ at a particular value of the relative velocity leads direct to b_{max} at the same relative velocity,

$$\sigma_{total}(v) = \pi b_{max}^2 \tag{7.33}$$

and collisions occur for values of b between $b = 0$ and $b = b_{max}$.

The same procedure yields $\sigma_{total}(v)$ and σ_{total} for beam B.

However, there are many values of b between $b = 0$ and $b = b_{max}$ which lead to collisions, and if any more detailed information is required it is necessary to study the angular scattering.

7.5.1 Angular scattering

These are important experiments since they lead to values of b corresponding to particular angles of scattering θ. The scattering into the small range of angles $d\theta$ around any given angle θ can be detected using a mass spectrometer to estimate the amount of scattered material hitting the "inside surface of the experimental sphere" described by the detector. In effect, what happens is that this sphere has been chopped up into slices of thickness dl, Figure 7.10, each slice corresponding to a given angle.

The initial intensity of the beam and the intensity into angle θ can, therefore, both be found, and these are linked by a Beer's Law type relation similar to that derived for total scattering. A value of the experimental cross section $\sigma^*(\theta, v)$ follows for the given relative velocity at which the experiment is carried out. Again an experimental cross section independent of velocity $\sigma^*(\theta)$ is found by integrating over the distribution of velocities used.

$$\sigma^*(\theta) = \int_0^\infty \sigma^*(\theta, v) f(v) dv. \tag{7.34}$$

Likewise, values of $\sigma^*(\theta, v)$ and $\sigma^*(\theta)$ can be found for beam B.

7.5.2 Measured cross sectional areas $\sigma^*(\theta, v)$ and differential cross sectional areas $\sigma(\theta, v)$

Different impact parameters give different angles of scattering; scattering into an angle θ_1 results from impact parameter b_1, angle θ_2 from b_2 and so on. The smaller the impact parameter, the larger is the angle of scattering. Because the impact parameter, b, is a microscopic quantity compared to the macroscopic quantity, angular scattering into angle θ, all impact parameters are coalesced to a point when a diagram showing possible angles of scattering is drawn, Figure 7.17.

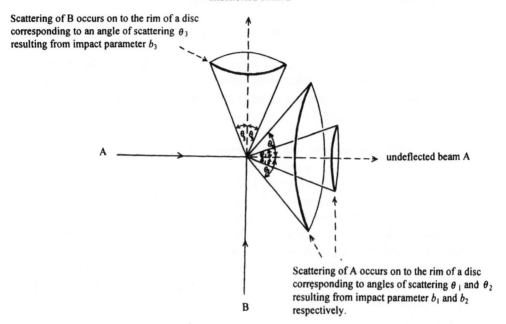

undeflected beam B

Scattering of B occurs on to the rim of a disc
corresponding to an angle of scattering θ_3
resulting from impact parameter b_3

A

undeflected beam A

Scattering of A occurs on to the rim of a disc
corresponding to angles of scattering θ_1 and θ_2
resulting from impact parameter b_1 and b_2
respectively.

B

Figure 7.17   Scattering from various impact parameters leads to scattering into
corresponding angles.

What is *actually* measured is scattering into a very small range of angles $d\theta$ around the
angle θ, and this results from a small range of impact parameters db around the impact
parameter b, Figure 7.18.

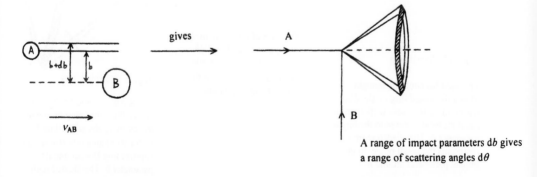

Scattering on the rim of disc of width
resulting from a small range of angles $d\theta$

gives   A

A

b+db   b

B

B

v_{AB}

A range of impact parameters db gives
a range of scattering angles $d\theta$

**Figure 7.18 Scattering into a range of angles $d\theta$ results from a range of impact parameters
db.**

The **measured** cross section $\sigma^*(\theta, v)$ is related to the **differential** cross section $\sigma(\theta, v)$ normally quoted in tabulated results at a given velocity, v.

$$\sigma^*(\theta, v) = \sigma(\theta, v)d\Omega \tag{7.35}$$

where $d\Omega$ takes charge of the small range of angles which are inherent in any experimental measurements. When most tabulated data are given as the differential cross section $\sigma(\theta, v)$ it is necessary to interconvert $\sigma(\theta, v)$ and $\sigma^*(\theta, v)$ using $d\Omega$. $d\Omega$ will be derived later, Section 7.6.

7.6 Detailed derivation of the relation between θ and b

The amount of scattering into a small range of angles $d\theta$ around any given angle θ is a direct result of collisions occurring with a small range of impact parameters, db, around the value of b which corresponds to the particular angle θ under consideration. It is possible to calculate the total amount of scattering resulting from all values of the impact parameter between $b = 0$, corresponding to a head on collision with total rebound backwards, and $b = b_{max}$. Likewise it is possible to calculate the total amount of scattering into all possible angles θ between $\theta = 180°$, corresponding to a head on collision, and $\theta = 0$, and from these two quantities derive a relation between b and θ. This is given below.

Figure 7.19 shows the geometrical relation between a range of impact parameters db and the resulting range of scattering angles.

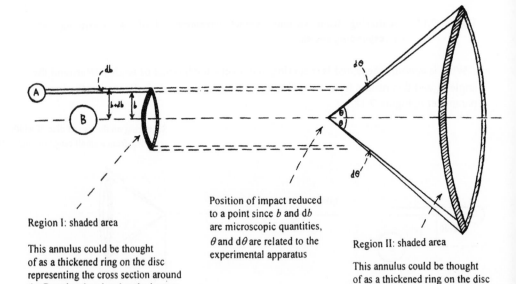

Region I: shaded area

This annulus could be thought of as a thickened ring on the disc representing the cross section around the B molecule related to the impact parameter b. The shaded region represents the small range db in impact parameter around b.

Position of impact reduced to a point since b and db are microscopic quantities, θ and $d\theta$ are related to the experimental apparatus

Region II: shaded area

This annulus could be thought of as a thickened ring on the disc representing the scatteringof the A molecule at an angle θ as a result of approaching B with impact parameter b. The shaded region represents the small range in angle $d\theta$ around θ result from the

Figure 7.19 Range of angles of scattering and range of values of b. range db in impact parameter b.

Scattering into a small range of angles $d\theta$ results in an annulus, or rim on the disc of scattering of definite thickness given by the shaded filled-in region of scattering in region II. This scattering results from a range of impact parameters db given by the filled-in region I.

Region II in Figure 7.19 is a result of scattering between the two cones of scattering of angle θ and angle $\theta + d\theta$, and it specifies a solid angle defined by the geometrical relation

$$d\Omega = \frac{\text{area of shaded region II}}{R^2}$$

(7.36)

where R is the slant height of the cone, Figure 7.20, and A represents the point of impact which is fixed geometrically.

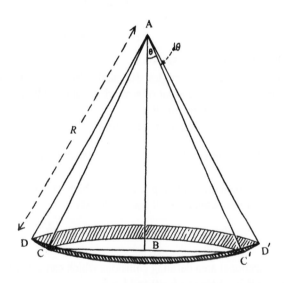

Figure 7.20 Scattering in terms of a solid angle $d\Omega$.

A geometrical argument is used to find the area of the shaded region II.

$$\text{the area of shaded region II} = \text{length of the perimeter} \times CD$$
$$= 2\pi BC \times CD.$$

(7.37)

But $CD = Rd\theta$ (by definition)

and $BC = R\sin\theta$

and so

$$\text{the area of shaded region II} = 2\pi R^2 \sin\theta d\theta$$
$$= R^2 d\Omega,$$

(7.38)

so that $$d\Omega \quad = 2\pi \sin \theta d\theta. \tag{7.39}$$

The amount of scattering, due to the range in angles $d\theta$ around θ, which results from a range in impact parameters db around b, is proportional to the quantity $d\Omega$, from which it follows that

the amount of scattering
into a range of angles $\propto 2\pi \sin \theta d\theta.$
$d\theta$ around θ

$$\tag{7.40}$$

The experimental cross section $\overset{\rightarrow}{\sigma}*(\theta, v)$ is also proportional to $d\Omega$, Section 7.5.2,

$$\sigma*(\theta, v) \propto d\Omega \tag{7.41}$$

$$\propto 2\pi \sin \theta d\theta$$

$$\tag{7.42}$$

$$= \sigma(\theta, v)2\pi \sin \theta d\theta$$

where $\sigma(\theta, v)$ is the differential cross section.

The small range of scattering angles $d\theta$ results from a small range in impact parameters db around b, and the amount of scattering which will result from this small range can again be calculated geometrically.

The area of region I, Figure 7.19, is found as follows:

the area of the outer disc $= \pi(b + db)^2$

and this *is* the cross sectional area defining collisions which can occur if the radius of the disc around molecule B lies between $b = 0$ and $b = b + db$

the area of the inner disc $= \pi b^2$

and this *is* the cross sectional area defining collisions which can occur if the radius of the disc around molecule B lies between $b = 0$ and $b = b$.

Scattering resulting from impact parameters lying between b and $b + db$ is given by the difference in the two cross sectional areas, and this can be formulated in terms of the area of region I.

area of region I $= \pi(b + db)^2 - \pi b^2$

$$= 2\pi b db \tag{7.43}$$

if db^2 is to be considered negligible with respect to $2\pi\, b\, db$.

Hence scattering due to a range db in the impact parameter b is proportional to $2\pi\, b\, db$, and, as argued above, the area $2\pi\, b\, db$ is the difference in the two cross sectional areas, and must therefore be the experimental cross section $\sigma*(\theta, v)$ for scattering into a range of angles $d\theta$ around θ.

$$\therefore \quad \sigma*(\theta, v) = 2\pi b db. \tag{7.44}$$

But it has already been shown, Equation 7.42 , that

$$\sigma*(\theta, v) = \sigma(\theta, v) 2\pi \sin \theta d\theta \qquad (7.45)$$

$$\sigma(\theta, v) = \frac{bdb}{\sin \theta d\theta}$$

or $\sigma(\theta, v) \sin \theta d\theta = bdb.$

$$(7.46)$$

Throughout this derivation the full statement, "scattering into a range of angles $d\theta$ around θ results from a range of impact parameters db around b" has been used. It is, however, general practice to condense this to a statement reading, "scattering into an angle θ results from an impact parameter b", but it must always be remembered that experiment can only measure scattering into a small range of angles, and *not* scattering into a precisely determined unique value of θ. Likewise, scattering does not occur from one precisely determined impact parameter.

It has been shown above, Equation 7.46 that

$$\sigma(\theta, v) \sin \theta d\theta = bdb.$$

Unfortunately b db cannot be evaluated directly, but the integral over all values of b from $b = 0$ to $b = b$ can be found. This requires that $\sigma(\theta, v) \sin \theta \, d\theta$ must also be integrated over the corresponding limits for the angle of scattering, that is from $\theta = \pi$, corresponding to the head-on collision $b = 0$, to $\theta = \theta$, corresponding to $b = b$.

$$\int_{\pi}^{\theta} \sigma(\theta, v) \sin \theta d\theta = \int_{0}^{b} bdb \qquad (7.47)$$

$$= \frac{b^2}{2}. \qquad (7.48)$$

The left hand integral cannot be integrated explicitly, but the integral corresponds to the area under the curve of $\sigma(\theta, v) \sin \theta$ against θ, Figure 7.21. All the quantities required for this graph are available from experiment, and the area between π and any angle can be found, and

$$\text{the area between } \pi \text{ and } \theta = \frac{b^2}{2}.$$

$$(7.49)$$

Hence b can be found for any θ.

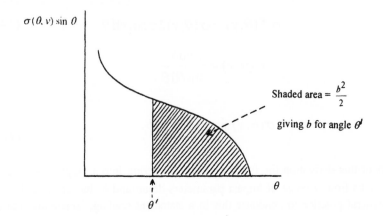

Figure 7.21   Graphical determination of b from the graph of $\sigma(\theta, v) \sin \theta$ against θ.

This is a relatively easy, direct and accurate route to values of b, and is also the only experimental method available. There are several important uses for knowledge of impact parameters.

Differential cross sections $\sigma(\theta, v)$ can be obtained from experiments where the colliding molecules are initially in specified quantum states and end up in a variety of quantum states after collision. During a collision energy is transferred from one molecule to another, and determination of the quantum states of the scattered molecules allows the type of energy transfer process to be identified. Angular scattering experiments lead, therefore, to values of the impact parameter for energy transfer between two specific quantum states before and after collision.

The major use of these experimental impact parameter values is for calculating intermolecular potential energies from which potential energy surfaces for energy transfer can be accurately derived, and, in particular, the region of the surface corresponding to short intermolecular distances can be accurately described. It is at these short distances that the quantum calculations are at their least accurate. If angular scattering for collisions with energy transfer between specific quantum states is found, then the derived potential energy surfaces become very detailed and useful.

Figure 7.15, Section 7.41, describes the graph of θ against b, and this shows clearly the influence of attractions and repulsions on collision.

The total cross section $\sigma_{total}(v)$, Section 7.4.2, is the sum of scattering into a range of angles $d\theta$ around θ over all angles θ.

$$\sigma_{total}(v) = \sum_{\text{all angles}} \sigma^*(\theta, v) \tag{7.50}$$

$$= \sum_{\text{all angles}} \sigma(\theta, v) d\Omega \tag{7.51}$$

$$= \int_0^{4\pi} \sigma(\theta, v)d\Omega. \tag{7.52}$$

7.7 Derivation of the rate of collision for energy transfer processes

(a) The derivation is the same as that given for simple collision theory, only $\pi(r_A + r_B)^2$ is replaced by πb^2_{max}.

The rate of collision   =   the total number of collisions per unit time per unit volume

$$= \mathcal{Z}_{AB}$$

$$= \sigma_{total}(v)N_A N_B v \tag{7.53}$$

and $$k(v) = \sigma_{total}(v)v. \tag{7.54}$$

A summation over the distribution of velocities gives a rate constant independent of v

$$k = \int_0^\infty \sigma_{total}(v)vf(v)dv. \tag{7.55}$$

(b) The arguments are similar where scattering into angle θ is being measured.

The rate of collision into angle $\theta = \sigma*(\theta, v)N_A N_B v$ $\qquad$ (7.56)

$$k(\theta, v) = \sigma*(\theta, v)v \tag{7.57}$$

giving $$k(\theta) = \int_0^\infty \sigma*(\theta, v)vf(v)dv. \tag{7.58}$$

(c) The rate of collision between molecules in specific quantum states is

$$k(sp.q.s., \theta, v) = \sigma*(sp.q.s., \theta, v)v \tag{7.59}$$

and $$k(sp.q.s., \theta) = \int_0^\infty \sigma*(sp.q.s., \theta, v)vf(v)dv. \tag{7.60}$$

7.8 Dependence of the total cross section σ_{total} on velocity

Experiments over a range of relative velocities and translational energies give the following graph, Figure 7.22.

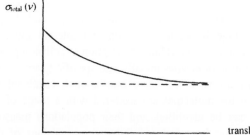

Figure 7.22 Dependence of the total cross-section on velocity.

This graph shows immediately that molecules cannot be treated as hard spheres. If they were hard spheres there would be no interactions and no scattering for impact parameters greater than $r_A + r_B$, and the cross section would remain constant at $\pi(r_A + r_B)^2$ giving a horizontal line. Experiment shows that this is not so.

The shape of the graph can be discussed purely qualitatively. If the molecules move together very slowly with low translational energies then the forces of interaction have long enough to have an effect, and a collision can occur, **even with** a large impact parameter. But if the molecules move together quickly with high translational energies they will fly past each other without interaction, and there will be no collision. This is especially so at large impact parameters. The lower the value of the impact parameter the more does the effect of a small b outweigh the "flying past effect", so that the cross section is large at low translational energies, decreasing at high energies, until it reaches the hard spheres value.

The hard spheres value is physically highly unrealistic, and the actual interaction is governed by attractions and repulsions. At large distances attraction is dominant resulting in a cross section larger than the hard spheres value at low and moderate translational energies. At small distances repulsion is dominant, and this leads to a cross section approaching the hard sphere value at high translational energies. At intermediate distances repulsion and attraction are both significant so that the cross section takes an intermediate value. A large range of models can describe this behaviour, and molecular beam experiments can test these models.

7.8.1 Composite nature of σ_{total}

Scattering can result from a collision resulting in energy transfer, or from a reactive collision. Energy transfer can be between translation, rotation and vibration. Contributions to the total scattering can be classified as

1. elastic scattering where the only exchange of energy is between translation and translation, giving a cross section $\sigma_{elastic}(v)$
2. non elastic scattering where there is exchange of energy between translation and the internal modes, giving $\sigma_{non\ elastic}(v)$ which is itself composite, made up from translation/rotation, rotation/vibration, vibration/vibration, and other such combinations of energy transfer.

$$\sigma_{non\ elastic}(v) = \sigma_{trans/rot}(v) + \sigma_{rot/vib}(v) + \sigma_{vib/vib}(v) + \ldots \qquad (7.61)$$

3. reactive scattering gives a cross section $\sigma_R(v)$ discussed in Sections 7.10.2-7.10.5.

Each of these energy exchanges can be studied in turn, and, in particular, those describing energy exchange between translation and internal modes, and between internal modes result in a much more refined description of collision processes.

To do this requires putting molecules initially into specific quantum states so that the molecules in the beam all have the same translational, rotational and vibrational energies. After collision the molecules are scattered with a range of quantum states. These quantum states can be identified, and their populations measured using laser absorption spectroscopy and laser induced fluorescence. Values of the translational energies can be found using the time of flight mass spectrometer. Values of the translational energies and the rotational and vibrational quantum numbers before and after collision define a specific energy transfer process. Cross sections can be found for all these energy transfers between pairs of specific quantum states before and after collision. If angular scattering is also studied then values of the angular cross section can be found for these state to state energy transfers.

Only a few collision processes have been studied in this amount of detail, but the potential is there. Further, if the populations and velocities of all product states are known, then the change in the distribution of energies as a result of collision can be found.

7.8.2 Change in the distribution of energies on collision
The initial quantum states in beams A and B are selected.

Beam A
translational energy = $\frac{1}{2}m_A v_A^2$
rotational energy = $J_A(J_A+1)Bh$
vibrational energy = $(v_A +\frac{1}{2})hv_{oA}$

Beam B
translational energy = $\frac{1}{2}m_B v_B^2$
rotational energy = $J_B(J_B+1)Bh$
vibrational energy = $(v_B +\frac{1}{2})hv_{oB}$

After collision the quantum states and their populations can be found for all scattered molecules of A and B, along with their translational energies.

Beam A
translational energy = $\frac{1}{2}m_A v'^2_A$
rotational energy = $J_A'(J_A'+1)Bh$
vibrational energy = $(v_A'+\frac{1}{2})hv_{oA}$

Beam B
translational energy = $\frac{1}{2}m_B v'^2_B$
rotational energy = $J_B'(J_B' + 1)Bh$
vibrational energy = $(v_B'+\frac{1}{2})hv_{oB}$

The total energy is conserved on collision so-that

$$\frac{1}{2}m_A v_A^2 + J_A(J_A + 1)Bh + (v_A +\frac{1}{2})hv_{0A} +\frac{1}{2}m_B v_B^2 + J_B(J_B +1)Bh$$

$$+ (v_B +\frac{1}{2})hv_{0B}$$

$$=\frac{1}{2}m_A v_A'^2 + J_A'(J_A' +1)B'h + (v_A' +\frac{1}{2})hv_{0A} +\frac{1}{2}m_B v_B'^2$$

$$+ J_B'(J_B' +1)B'h + (v_B' +\frac{1}{2})hv_{0B} \qquad (7.62)$$

and $\Delta\varepsilon_{\text{trans}}$, $\Delta\varepsilon_{\text{rot}}$ and $\Delta\varepsilon_{\text{vib}}$ can be found for each of A and B.

If one of the final quantum numbers or velocities cannot be found, then the above equation will allow it to be found by calculation. For instance, if the final velocities of scattered A and B in specific rotational and vibrational states cannot be found from the time of flight mass spectrometer, then the above equation can be used to calculate $\frac{1}{2}m_A v'^2_B + \frac{1}{2}m_B v'^2_B$. Likewise, if the initial and final velocities are known then it is still possible to infer details of internal energy changes.

As stated earlier if the populations of the vibrational and rotational states of the reactants and products can be found, then the finer details of energy transfer and the redistribution of energy on collision can be found. This gives vital information on the activation and deactivation processes in a chemical reaction, and similar studies on reactive collisions gives corresponding detailed information about the reaction process (Sections 7.10 onwards).

7.8.3 Selection of the energy states of the colliding molecules

A rather unspecific experiment would use beams without any selection where equilibrium distributions of translational, rotational and vibrational energies are present. Most work is carried out in the ground electronic state, though there is no restriction, in principle, to this state.

Supersonic and seeded beams have well defined velocities and need no further selection. Such beams have high intensities, and a wide range of velocities can be selected.

Molecules are generally in their ground vibrational state and need no further selection. However, they are likely to have an equilibrium distribution of rotational energies, with a large number of rotational states populated. For polar molecules it is possible to select one of these states by using focused electric fields. It is easier, however, to selectively populate a given rotational state of the ground vibrational state using a maser of the appropriate frequency. The high intensity of the maser results in a high population of the selected state.

Vibrational states other than the ground state, must be selected by laser, and this will put the molecules into a specific rotational state in a specific vibrational state. And again, because of the high intensity of the laser, a relatively high population of the required state will be generated. With the range of lasers now available a large number of rotational and vibrational states can be populated.

Molecular beam experiments can measure the total cross sectional area, πb^2_{max}, for specific state to state processes. More refined experiments will measure the cross section for scattering into an angle θ, and in doing so furnish the impact parameters for specific state to state processes.

More definitive experiments keep the velocity and rotational energy constant, and determine the effect of vibrational energy on the cross section, or likewise determine the effect of rotational energy for experiments with constant velocity and vibrational energy, and so forth. Similar experiments on reactive scattering determine threshold energies for reaction, Sections 7.10.1 to 7.10.3, and give details of translational and vibrational enhancement of reaction which have proved so important in looking at details on potential energy surfaces, Chapter 6.

Selection of states for each colliding beam results in a considerable reduction of the beam, and consequent reduction in the intensity of the scattered molecules. This is particularly so if translational, rotational and vibrational energy is selected in the colliding beams, and if transfer is into specific quantum states of the scattered beams. Developing techniques of detection should alleviate this problem.

7.8.4 Experimentally derived quantities in state to state energy transfer

Once a set of initial and final quantum states are chosen it has been shown that the following quantities can be found for each state to state process:

1. the total cross sections $\sigma_{total}(v)$ and σ_{total}

2. the angular cross sections $\sigma^*(\theta, v)$ and $\sigma^*(\theta)$, together with the differential cross sections $\sigma(\theta, v)$, $\sigma(\theta)$
3. the rate constants $k_{\text{total}}(v)$, k_{total}, and the individual rate constants $k(\theta, v)$ and $k(\theta)$
4. the detailed rate constants $k(\theta, v, \text{sp.q.s to sp.q.s})$ and $k(\theta, \text{sp.q.s to sp.q.s})$

Although these detailed experiments require time to obtain the results, they result in a vastly deeper understanding of collision processes than do the simple total cross sections.

7.8.5 Resonant and non-resonant energy changes
This can be illustrated by considering the following collision process

$$A + B \xrightarrow[\text{transfer}]{\text{energy}} A + B$$

where only the vibrational quantum numbers v_A and v_B change. The question can be asked, "is this process purely vibration to vibration, or is it vibration to vibration plus vibration to translation, or is it vibration to vibration plus translation to vibration?"

If the vibrational energy change in A equals the vibrational energy change in B, then the energy change is exactly resonant, and there will be no consequent change in translational energy. The energy change will be purely vibrational to vibrational energy transfer, able to be confirmed by observation of the initial and final translational energies being equal.

But if the vibrational energy change in A is not equal to the vibrational energy change in B, then the exchange is non-resonant, and there will be a consequent change in translational energy which can be observed. The change will, therefore, be a vibrational to vibrational energy change with simultaneous vibrational translational energy transfer.

The same reasoning applies to rotational energy transfer, or vibrational-rotational transfer.

7.8.6 A typical experiment studying the kinetics of energy transfer
For simplicity, energy transfer is considered between a diatomic molecule with one vibration, and with quantum numbers, v_A and J_A, and an atom which has no rotations or vibrations, Table 7.1.

Table 7.1 Changes in quantum numbers and energy transfer processes in collisions

Colliding molecules		Scattered molecules		Rate constant for state to state transfer	type of energy transfer
A	B	A	B		
diatomic molecule	atom	diatomic molecule	atom		
$v=3, J=7$	-	$v=3, J=6$	-	$k(3376)$	rot-trans
	-	$v=3, J=5$	-	$k(3375)$	rot-trans
	-	$v=3, J=8$	-	$k(3378)$	trans-rot
	-	$v=3, J=9$	-	$k(3379)$	trans-rot

$v=3, J=7$	-	$v=2, J=7$	-	$k(3277)$	vib-trans
	-	$v=2, J=6$	-	$k(3276)$	vib and rot-trans
	-	$v=2, J=5$	-	$k(3275)$	vib and rot-trans
	-	$v=2, J=8$	-	$k(3278)$	vib-rot and trans
	-	$v=2, J=9$	-	$k(3279)$	vib-rot and trans
$v=3, J=7$	-	$v=4, J=7$	-	$k(3477)$	trans-vib
	-	$v=4, J=6$	-	$k(3476)$	trans and rot-vib
	-	$v=4, J=5$	-	$k(3475)$	trans and rot-vib
	-	$v=4, J=8$	-	$k(3478)$	trans-vib and rot
	-	$v=4, J=9$	-	$k(3479)$	trans-vib and rot

The initial quantum numbers for molecule A are v_A and J_A and the scattered molecules are studied to find all the product v_A and J_A values. Cross sections are found for scattering into the observed quantum states of A, and the corresponding rate constants for each state to state energy transfer process can be calculated. From experiments like these it is possible to build up data on total cross sections, differential cross sections and rate constants for energy transfer between specific quantum states. By comparing the quantum numbers v_A and J_A for molecule A before collision with those after collision it is possible to work out the energy transfers given in the last column of Table 7.1.

Molecular beam techniques are easily the simplest, most straight-forward and most accurate experimental tool for obtaining such data. When this is combined with corresponding reactive cross sections and rate constants the microscopic rate constants for all activation, deactivation and reaction steps can, in principle, be inferred for a given reaction, and the overall rate of any elementary process can be found in terms of these microscopic quantities.

7.9 Scattering diagrams with respect to the centre of mass
So far all the diagrams have been given in terms of the laboratory situation, for instance angles of scattering have been drawn in with respect to what is actually measured in the molecular beam apparatus. However, all molecular beam results are tabulated in terms of scattering angles γ which are referred to the centre of mass of the colliding particles. Inferential results such as the contour diagrams in Chapter 10 are all drawn using these centre of mass scattering angles γ.

The observed scattering cones given as θ for A and B have to be converted into a scattering cone with respect to the centre of mass. Likewise angular intensities of the scattered molecules have to be converted.

One of the benefits of this conversion is that there is conversion of *two* cones of scattering θ_A, θ_B into *one* cone of scattering with angle γ with respect to the centre of mass. This makes the subsequent manipulation of the experimental data very much easier.

7.9.1 Forward scattering of both A and B
Scattering with respect to the laboratory apparatus has *two* totally distinct cones of scattering, A having angle θ_A and B having angle θ_B, Figure 7.23.

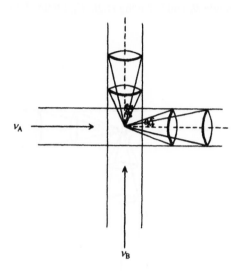

Figure 7.23 Cones of forward scattering for beams A and B in terms of the laboratory angle of scattering.

This converts to *one* double cone with a single scattering angle with respect to the centre of mass, Figure 7.24.

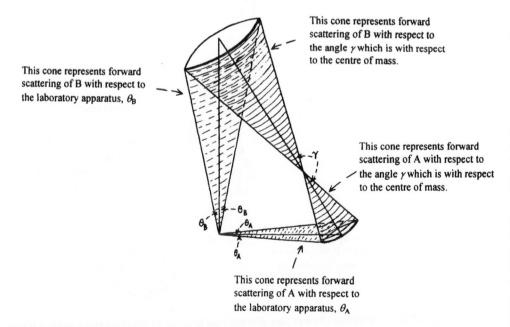

This cone represents forward scattering of B with respect to the angle γ which is with respect to the centre of mass.

This cone represents forward scattering of B with respect to the laboratory apparatus, θ_B

This cone represents forward scattering of A with respect to the angle γ which is with respect to the centre of mass.

This cone represents forward scattering of A with respect to the laboratory apparatus, θ_A

Figure 7.24 Conversion of two cones of scattering in terms of the laboratory angle θ into a single cone of scattering in terms of the centre of mass angle γ.

7.9.2 Forward scattering of A, backward scattering of B

Again scattering with respect to the laboratory apparatus gives *two* totally distinct cones of scattering, A having angle θ_A and B having angle θ_B, Figure 7.25.

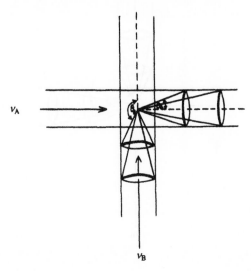

Figure 7.25 Cones of forward scattering of A and backward scattering of B in terms of the laboratory angle of scattering.

Dealing *with* one double cone with *one* angle makes all the subsequent analysis easier, Figure 7.26.

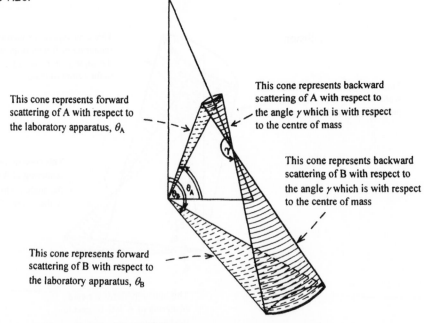

This cone represents forward scattering of A with respect to the laboratory apparatus, θ_A

This cone represents backward scattering of A with respect to the angle γ which is with respect to the centre of mass

This cone represents backward scattering of B with respect to the angle γ which is with respect to the centre of mass

This cone represents forward scattering of B with respect to the laboratory apparatus, θ_B

Figure 7.26 Conversion of two cones of scattering in terms of the laboratory angle θ into a single cone of scattering in terms of the centre of mass angle γ.

7.10 Reactive collisions
Scattering experiments can be carried out for reacting gases

$$A + B \quad \rightarrow \quad C + D$$

and two types of scattering can be measured. The total scattering due to non-reactive and reactive collisions can be found from the drop in intensity of A or B. Scattering due to reaction can be found by measuring the scattering of products C or D. The difference between these two experiments would give the scattering due to non-reactive collisions. Total and differential cross sections can be measured for all collisions, while total and differential cross sections for forming the scattered products C and D are cross sections for reactive collisions only. If the quantum states of the reactants have been selected then more information will emerge if the quantum states of scattered A and B, and the quantum states for scattered products C and D are found.

7.10.1 The cross section for reaction
In reactive scattering all molecules which have reacted must also have collided, and the definition of a collision remains the same as for energy transfer collisions, Figure 7.27.

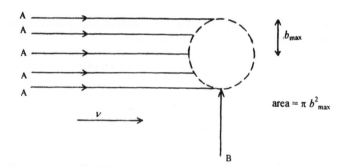

Figure 7.27 Definition of a collision in reactive scattering.

All molecules of A which hit the disc of radius b_{max} and cross section πb^2_{max} around B will collide with B, and the molecules which do so will be made up of A molecules which simply collide with B and do not react, as well as A molecules which do react on collision with B. The cross section for all collisions, reactive plus non-reactive, is πb^2_{max}.

All molecules which react have collided, but not all molecules which collide will react. This is a consequence of the quantum mechanical nature of reaction where bonds are formed and broken. Classical dynamics, on which most molecular beam calculations are based, adequately describes the trajectories for non-reactive scattering. However, the movement of two molecules undergoing reaction is inherently quantum mechanical, and this means dealing with probabilities. If two molecules actually collide there is only a certain probability that they will react.

To take this into account the reaction probability $P(b)$ is defined to be the fraction of collisions with impact parameter b which lead to reaction, and so $P(b)$ lies between zero and unity. When $P(b)$ is zero then none of the collisions result in reaction, if $P(b)$ is unity then all of the collisions lead to reaction, and at intermediate values of $P(b)$ some collisions lead to reaction. $P(b)$ is significant for values of b where intermolecular forces are effective, but falls to zero for values of b greater than b_{max}.

There must be no suggestion that the statement "not all collisions lead to reaction" means that some molecules are insufficiently energised. $P(b)$ is taking account of a quantum mechanical effect on the trajectories, and has nothing to do with activation.

7.10.2 Calculation of reactive cross sections

The cross section, $\overset{*}{\sigma}(\theta, v)$ for scattering into a range of angles $d\theta$ around angle θ from a range of impact parameters db around b and at relative velocity v is given by Equation 7.44

$$\overset{*}{\sigma}_{\text{all collisions}}(\theta, v) = 2\pi b db. \tag{7.63}$$

The total scattering for all collisions is

$$\overset{*}{\sigma}_{\text{all collisions}}(v) = \int_0^{b_{max}} 2\pi b db \tag{7.64}$$

$$= \pi b_{max}^2. \tag{7.65}$$

When reactive collisions are considered the corresponding quantities must include the probability of reaction, $P(b)$, as a result of collision

$$\overset{*}{\sigma}_{\substack{\text{reactive} \\ \text{collisions}}}(\theta, v) = 2\pi b P(b) db, \tag{7.66}$$

$$\overset{*}{\sigma}_{\substack{\text{total reactive} \\ \text{collisions}}}(v) = \int_0^{b_{max}} 2\pi b P(b) db. \tag{7.67}$$

This last equation can only be integrated if $P(b)$ is independent of b, which is rare, or by numerical integration if $P(b)$ and b are known.

7.10.3 Determination of $P(b)$

$P(b)$ is defined as

$$P(b) = \frac{\text{reactive collisions into angle } \theta}{\text{all collisions into angle } \theta} \tag{7.68}$$

$$= \frac{\overset{*}{\sigma}_R(\theta, v)}{\overset{*}{\sigma}_{\text{all collisions}}(\theta, v)}. \tag{7.69}$$

Both these quantities can be measured as described previously.

Determinations of b

What is required here is the cross section for *all* collisions into a series of angles θ. A graph of

$$\sigma_{\text{all collisions}}(\theta, v) \sin \theta \text{ against } \theta$$

is drawn, and the area under the graph between any angle θ' and π is $b^2/2$, where b is the value of the impact parameter resulting in scattering into angle θ', Section 7.6 and Figure 7.21.

Combination of $P(b)$ and b

From values: (i) b for various θ

                (ii) $P(b)$ for various θ

the value of $P(b)$ for corresponding values of b can be found, and a graph of $P(b)$ against b can be drawn, Figure 7.28. For values of b greater than b_{max}, there are no collisions and $P(b) = 0$, while at $b = 0$ all collisions are successful in causing reaction and $P(b) = 1$

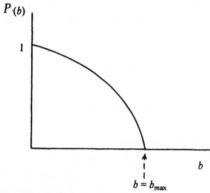

Figure 7.28 Graph of $P(b)$ against b.

7.10.4 Total rate constants for reaction

A derivation exactly analogous to that for collision theory, Sections 7.1.2 to 7.1.4, gives an expression for the rate of reaction defined to be the number of reactive collisions per unit volume per unit time.

$$\text{rate of reaction} = Z_{AB}$$

$$= \sigma_{R\,total}(v)vN_A N_B \tag{7.70}$$

$$\text{giving} \quad k = \sigma_{R\,total}(v)v. \tag{7.71}$$

A rate constant independent of velocity is found by summing over all velocities given by the distribution of velocities $f(v)\, dv$

$$k = \int_0^\infty \sigma_{R\,total}(v)vf(v)dv. \tag{7.72}$$

Often a Maxwell-Boltzmann distribution is used giving

$$k = \left(\frac{1}{\pi\mu}\right)^{1/2}\left(\frac{2}{kT}\right)^{3/2}\int_0^\infty \sigma_{R\,total}(\varepsilon)\exp\left(-\frac{\varepsilon}{kT}\right)d\varepsilon \tag{7.73}$$

where ε is the kinetic energy $= \frac{1}{2} mv^2$

7.10.5 Rate constants for reactive angular scattering

The cross section, $\sigma^*_R (\theta, v)$, for reactive scattering into angle θ, allows the rate constant for the reaction for these conditions to be found.

$$k(\theta,v) = \sigma_R^*(\theta,v)v \qquad (7.74)$$

$$k(\theta) = \int_0^\infty \sigma_R^*(\theta,v)vf(v)dv. \qquad (7.75)$$

Values of θ correspond to particular values of b and $P(b)$, and so rate constants for each value of b and $P(b)$ can be found. An enormous amount of information is contained in tabulated data listing

$$\sigma_R^* (\theta,v)\theta, k(\theta), b \text{ for each } \theta, P(b) \text{ for each } \theta$$

Even more information emerges if state to state rate constants are found.

7.10.6 General state to state rate constants

This requires measurement of the angular scattering from reaction between A in quantum state i, B in quantum state j, to give C in quantum state k and D in quantum state m. The corresponding cross section is $\sigma_R (\theta, v, i, j, k, m)$, and the rate constant is $k (\theta, v, i, j, k, m)$ so that

$$k(\theta,v,i,j,k,m) = \sigma_R (\theta,v,i,j,k,m)v, \qquad (7.76)$$

$$k(\theta,i,j,k,m) = \int_0^\infty \sigma_R (\theta,v,i,j,k,m)vf(v)dv. \qquad (7.77)$$

The experiments which can be carried out to determine these state to state rate constants are discussed along with other techniques in Chapter 8.

There are difficulties mainly due to the low intensity of the scattered beams of the products of reaction. This is particularly so when angular reactive scattering between specific quantum states is determined, and accurate measurements requires very sensitive detection. At present detection is the main limitation on observing state to state processes of energy transfer and reaction.

7.11.1 Effect of translational energy on the reaction cross section

When the total reaction cross section is plotted against translational energy, the reaction cross section is zero until a minimum energy is reached, whereupon it increases often to a broad plateau before decreasing again. Sometimes only an increase to the broad plateau is found.

The minimum energy ε_0 is called the ***threshold energy***, and the curve $\sigma_R(\varepsilon)$ against ε is called the ***excitation function***.

7.11.2 Two typical shapes of the excitation function
These are shown in Figures 7.29, and 7.30.

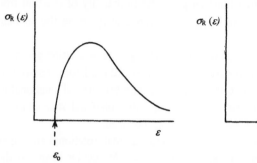

Figure 7.29 A typical excitation curve showing a maximum.   **Figure 7.30 A typical excitation function showing a plateau.**

The value of ε_0 can be measured precisely, and is the minimum relative kinetic energy for reaction to occur. It is not the bulk experimental activation energy, nor is it the theoretical activation energy calculated from the potential energy surface, but it is related to both.

Any point on the curve gives the probability of reaction for a given translational energy, and values of $k(\varepsilon)_{trans}$ over a range of ε_{trans} allows mathematical formulation of the reaction probability as a function of ε_{trans}.

7.11.3 Effect of vibrational energy on the reaction cross section
Specific initial vibrational states can be selected for one beam, and the reactive scattering for each determined for one given initial velocity, giving $\sigma_{vib, R}(\varepsilon_i)$, where ε_i represents the vibrational energy of quantum state i. The graph of $\sigma_{vib, R}(\varepsilon_i)$ against ε_{vib} sometimes shows a minimum threshold energy, or it may not, Figures 7.31 and 7.32.

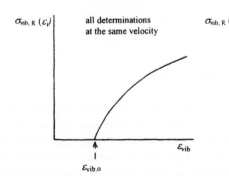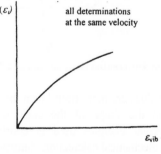

Figure 7.31 Graph of $\sigma_{vib, R}(\varepsilon_i)$ **against** against ε_{vib} for a reaction showing a threshold energy.

Figure 7.32 Graph of $\sigma_{vib, R}(\varepsilon_i)$ against ε_{vib} for a reaction with a zero threshold energy.

A minimum threshold energy indicates that a minimum vibrational energy is required before reaction can occur. Collision theory and transition state theory tacitly assume that the only requirement for reaction is that the kinetic energy of relative translational motion along the line of centres takes a critical value. A direct test of this assumption can now be made, and the molecular beam experiments show that for some reactions a critical minimum vibrational energy is also required.

Again a point on the excitation function gives the probability of reaction from any given vibrational state, and values of $k(\varepsilon)_{vib}$ over a range of ε_{vib} gives the mathematical formulation.

Similar experiments selecting initial rotational states give an excitation graph of $\sigma_{rot, R}$ (ε_j) against ε_j from which it can be shown whether a critical minimum rotational energy is required for reaction. Again some reactions do require this critical rotational energy, some do not. The probability of reaction from a specific rotational state can be found, and the mathematical relation between $k(\varepsilon_{rot})$ and ε_{rot} deduced.

The total reaction cross sections from specific translational, rotational and vibrational states lead to rate constants for very specific processes. Molecular beams is the only technique giving this information, and is the real basis for the study of state to state kinetics, see Chapter 8. It also gives information about the redistribution of energy on reaction.

7.11.4 Differential reaction cross sections
The differential cross section for reactive collisions is found for various translational energies, Figure 7.33.

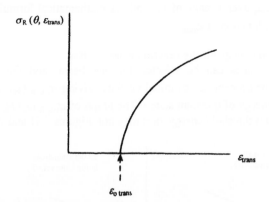

Figure 7.33 Differential cross-section as a function of energy.

These values are more useful than the total reaction cross sections as they give values of b, and the shape of the excitation functions gives extensive information on intermolecular energies used in constructing potential energy surfaces. Often the quantum mechanical calculations furnish the data for large intermolecular distances and molecular beam experiments fill in the details at short distances, particularly those around the critical configuration.

Graphs of $\sigma_R (\theta, \varepsilon_{vib})$, $\sigma_R (\sigma, \varepsilon_{rot})$ give similar information.

8

State to State Kinetics in Energy Transfer and Reaction

Chapter 1 Section 1.3 gives a master mechanism for all steps in elementary reactions. These are activation, deactivation and reaction. The energy transfer processes are characterised by composite rate constants which involve rate constants for all conceivable activation and deactivation steps between specified quantum states.

The energies associated with each quantum state are

$$\varepsilon_0, \varepsilon_1, \varepsilon_2, \varepsilon_3 \ldots\ldots \varepsilon_{n-1}, \varepsilon_n, \varepsilon_{n+1} \ldots\ldots .$$

where n is the **last** level below the critical energy and $n + 1$ is the **first** level above the critical energy.

Rate constants can be for
(a) "ladder climbing"; for energy transfer between adjacent levels or levels close together, and the amount of energy transferred is small,
(b) "strong collisions" where transfer is between non-adjacent widely spaced levels, and the amount of energy transferred is relatively large.

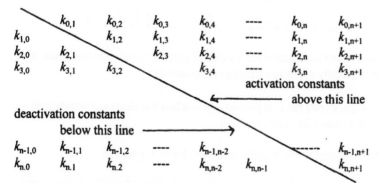

Modern work indicates that ladder climbing is the dominant mechanism for vibrational and rotational energy transfer.

The third step in the master mechanism is reaction. Here rate constants are required for reaction from specified quantum states in reactants to specified quantum states in the products.

This chapter describes methods of determining these microscopic state to state rate constants.

8.1 Use of molecular beams in generating state to state rate constants

At present, only a limited number of reactions have been studied in microscopic detail. The main drawback is that selection of initial quantum states reduces the intensity of the beams, and scanning the products for specific quantum states reduces the intensity yet further, so that detection of a specific quantum state and determination of its population become very difficult.

The type of molecular beam experiment depends on the information required, as shown below.

8.1.1 Energy transfer studies: non-reactive collisions

Example 1

The energy transfer process

$$A(v_A J_A \varepsilon_{trans\ A}) + B(v_B J_B \varepsilon_{trans\ B}) \rightarrow A(v_A' J_A' \varepsilon_{trans\ A}') + B(v_B' J_B' \varepsilon_{trans\ B}')$$

where all quantum states are completely specified before and after collision represents the ultimate aim in kinetics. In principle, molecular beams should furnish this data, but difficulties in detection and population determination are crucial factors in limiting the success of the experiment. In principle, the experiment could be as follows

1. Specified quantum states of A and B are generated by laser, and the populations determined by laser absorption spectroscopy or laser induced fluorescence.
2. These beams are shot at each other with a specified relative velocity.
3. The quantum states of A and B after the energy-transferring collision are determined by laser absorption spectroscopy and their populations likewise found using laser techniques.
4. From this a set of cross sections and rate constants can be found for energy transfer from *one* set of specified $v_A\ J_A\ \varepsilon_{trans\ A}$ and $v_B\ J_B\ \varepsilon_{trans\ B}$ into *all* possible quantum states $v_A'\ J_A'\ \varepsilon_{trans\ A}'$ and $v_B'\ J_B'\ \varepsilon_{trans\ B}'$.

$$\sigma(v_A J_A \varepsilon_{trans\ A}, v_B J_B \varepsilon_{trans\ B}; v_A' J_A' \varepsilon_{trans\ A}', v_B' J_B' \varepsilon_{trans\ B}')$$

5. These experiments can be repeated for a series of specific initial quantum states of A and B, so that ultimately all energy transfers are quantified.

Molecular beam experiments have, at present, not reached this degree of sophistication, and this remains a matter for future research.

From this, the rate constant at one specific relative velocity is given as

$$k(v) = \sigma(v; v_A J_A, v_B J_B; v_A' J_A', v_B' J_B') v \tag{8.1}$$

and the rate constant independent of velocity is

$$k = \int_0^\infty \sigma(v; v_A J_A, v_B J_B; v_A' J_A', v_B' J_B') v f(v) dv. \tag{8.2}$$

Example 2
A much less selective experiment where only the initial and final vibrational states of A are specified corresponds to transfer of vibrational energy to or from A. There may be consequent transfer of rotational and/or translational energy if the vibrational transfer is not resonant.

$$A(v_A, \text{ all } J_A \text{ and } \varepsilon_{\text{trans A}} \text{ unspecified})$$

$$+ B(\text{all } v_B J_B \text{ and } \varepsilon_{\text{trans B}} \text{ unspecified}) \rightarrow$$

$$A(v'_A \text{ all } J_A \text{ and } \varepsilon_{\text{trans A}} \text{ unspecified})$$

$$+ B(\text{all } v_B J_B \text{ and } \varepsilon_{\text{trans B}} \text{ unspecified}).$$

Since only the change in the vibrational state of A is being studied there is no need to select an initial quantum state of B or probe for specific quantum states of B after collision, so that only the total scattering of B need be found. Likewise there is no need to select or probe for specific rotational and translational states of A.

The molecular beam experiment could involve:

1. A specific vibrational state of A is selected by laser.
2. Beams of A and B are shot at each other with specified relative velocity.
3. All final vibrational states of A are probed and their populations found.
4. The total scattering of B is measured.
5. The experiment is repeated for other initial vibrational states of A.

Because selection is not highly specific, problems with intensities of initial and scattered beams are minimised. The cross section for each vibrational change in A is $\sigma(\varepsilon_{\text{trans A}} \, v_A, \, v_A')$, but the expression for the rate constant is very complex involving vibrational degeneracies, partition functions and sums over all rotational states.

Example 3
There are obviously processes which are intermediate in the detailed nature of the specification of the quantum states.

For example, there could be vibrational transfer between two molecules where the vibrational quantum numbers of A and B both change. If the vibrational change is resonant (Chapter 9, Section 9.3), there will be no consequent change in translational energy, and the change would be purely vibrational-vibrational. If the change is not resonant there is transfer to translation, and the change is vibration to vibration with accompanying vibrational-translational energy change. Likewise, there may, or may not, be accompanying rotational energy transfer.

The process would be

$$A(v_A J_A \varepsilon_{\text{trans A}}) + B(v_B J_B \varepsilon_{\text{trans B}}) \rightarrow A(v'_A J'_A \varepsilon'_{\text{trans A}}) + B(v'_B J_B \varepsilon'_{\text{trans B}})$$

where J_A and J_B are unspecified before and after collision.

The molecular beam experiment would be

1. Specific vibrational states of A and B are selected.
2. A and B are shot at each other with known relative velocity.
3. All vibrational states of A and B after collision are probed and their populations determined.
4. Comparison of final and initial velocities of A and B would tell whether the transfer is resonant or not.
5. This could be repeated for different initial vibrational states of A and B.

In general, the more detailed the specific quantum states involved the simpler is the expression for the cross section and rate constant, though there are then consequent problems of detection. When all rotational or vibrational states are allowed, then the calculation becomes very complex, but the problems of low intensities are minimised.

Each of these three examples of energy transfer are prototypes for all processes of energy transfer which can occur. Depending on the quantum states involved, the transfer will be either activation or deactivation. From this information it is, in principle, then possible to calculate k_1 for activation and k_{-1} for deactivation, Section 8.13.

8.1.2 Reaction: reactive collisions

In reactive collisions, scattering of the new product molecules is studied, and either the total scattering or scattering into specific angles is measured. In more detailed studies specific quantum states of the reactants are selected and the quantum states and the populations of the product molecules are probed. What has been said about non-reactive scattering can be taken over directly for reactive scattering; the essential difference between the two types of experiment is that new molecules are scattered in the reaction experiments.

For instance, reaction could be between A and B molecules in specific quantum states to give products C and D in specific quantum states,

$$A\left(v_A J_A \varepsilon_{\text{trans A}}\right) + B\left(v_B J_A \varepsilon_{\text{trans B}}\right) \rightarrow C\left(v_C J_C \varepsilon_{\text{trans C}}\right) + D\left(v_D J_D \varepsilon_{\text{trans D}}\right).$$

The cross section and rate constant are related, and for one given relative velocity, v

$$k\left(v; v_A J_A, v_B J_B; v_C J_C, v_D J_D\right) = \sigma\left(v; v_A J_A, v_B J_B; v_C J_C, v_D J_D\right) v \quad (8.3)$$

and gives a rate constant independent of velocity by integration over the appropriate distribution of velocities

$$k\left(v_A J_A, v_B J_B; v_C J_C, v_D J_D\right) = \int_0^\infty \sigma\left(v; v_A J_A, v_B J_B; v_C J_C, v_D J_D\right) v f(v) dv.$$

$$(8.4)$$

Each state to state rate constant determined in this way allows a calculation of k_s, the rate constant for reaction which appears in Equation 8.39

The calculation of an overall rate constant for reaction using these individual state to state rate constant for energy transfer and reaction is outlined in Section 8.13.

8.2 Relaxation techniques in state to state kinetics

A gas at equilibrium has a Maxwell-Boltzmann distribution of translational, rotational, vibrational and electronic energies. If this equilibrium is disturbed, the build up to the new required equilibrium distribution can be watched, and the rate of energy transfer found. The perturbation can be small, and the new equilibrium position only slightly away from the original equilibrium as in ultrasonics, temperature and pressure jump experiments, or it can be large and often very large, with the new equilibrium position far from the original as in shock tubes and flash photolysis. The movement to the new equilibrium is a *relaxation process*, and the time required for this change to occur when the displacement is small is the *relaxation time*.

Relaxation processes can be split into three types:
(i)   the single impulse displacements of temperature and pressure jumps, shock tubes and flash photolysis,
(ii) the pulsed displacements of pulsed shock tubes and pulsed radiolysis,
(iii) the periodic displacements of ultrasonic experiments.

In these processes, the initial transfer of energy is into translation, but this is then followed by typical relaxation processes such as translation to vibration, translation to rotation, and vibration to rotation or rotation to vibration, which occur until the new equilibrium is established.

8.2.1 Rates of energy transfer

In principle, all energy levels are involved and rate constants for each transfer can be formulated (Section 8.13).

In vibrational transfer, ladder climbing is involved, and harmonic oscillators are assumed, and rate constants for higher levels can be found in terms of transfer between the ground and first vibrational levels

$$k_{i+1,i} = (i+1)k_{10}. \tag{8.5}$$

This equation may be adequate for low-lying levels where vibrations approximate to harmonic oscillators, but it certainly is *not* valid at the high energies in which kineticists are interested.

A corresponding, but empirical, expression gives rate constants for rotational energy transfer between higher levels. This covers both ladder climbing and strong collision processes.

$$k_{J' \to J''} \propto \exp\{-ChB[J'(J'+1) - J''(J''+1)]\} \tag{8.6}$$

where C is a constant and B is the spectroscopic rotational constant.

8.3 Ultrasonic Techniques

This has been used to study fast reactions, but the technique can also be used to study energy transfer.

A gas or solution is displaced from an initial full Maxwell-Boltzmann distribution, and the system then relaxes to the new equilibrium position by collisional energy transfers. The relaxation time is the time at which the system has still to go $1/e = 0.368$ of the way to the new equilibrium. The speed at which these processes happen depends on the *frequency* of collisions, and on the *effectiveness* of collisions in transferring energy.

Interaction of a gas, or solution, with sound waves enables this process to be followed with ease. A sound wave corresponds to a sequence of alternating temperatures which corresponds to a sequence of alternating Maxwell-Boltzmann distributions. If the ground electronic state is being studied there will be a sequence of alternating Maxwell-Boltzmann distributions of translational, rotational and vibrational energies, Figures 8.1a, 8.1b, 8.1c.

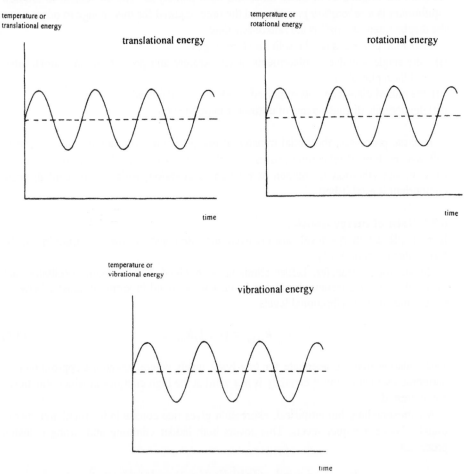

Figure 8.1a, b, c **Alternating Maxwell-Boltzmann distributions for translational, rotational, vibrational energies respectively.**

At *low* frequencies of the sound wave, the system adjusts in step with the alternations of temperature, and at every stage of the compression - rarefaction wave cycle the appropriate Maxwell-Boltzmann distributions are set up. This means that collisions are effective in transferring energy faster than the temperature is alternating, Figure 8.2a.

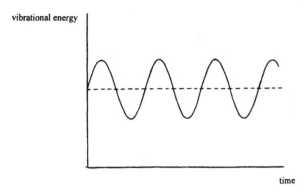

At low frequencies, vibrational energy transfer keeps the
vibrational energy changes in phase with the sound wave

**Figure 8.2a   Collisions are occurring fast enough, and are effective in transferring
vibrational energy to maintain a Maxwell-Boltzmann distribution.**

At **high** frequencies, complete readjustment of the system finally breaks down. Rotational energy transfer is much easier than vibrational energy transfer, and so the readjustments of vibrational energy drop out first.

When the period of the sound wave becomes comparable to the relaxation time for vibrational energy transfer, collisions will **just** fail to be effective in transferring translational energy of one molecule into vibrational energy in the other molecule involved in the collision. When this happens the vibrational energy remains almost constant throughout the cycle, instead of alternating, Figure 8.2b.

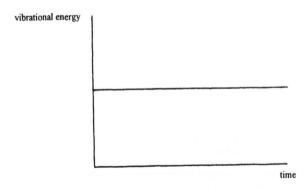

At higher frequencies, vibrational energy transfer cannot
keep in phase with the sound wave and no transfer
occurs, leaving the vibrational energy constant
throughout the disturbance created by the sound wave.

Figure 8.2b Collisions are not effective in transferring vibrational energy.

At **even higher** frequencies, collisions can no longer transfer rotational energy fast enough to keep up with the rapid alternations in temperature, and the rotational energy remains almost constant throughout the cycles, Figures 8.3a, 8.3b.

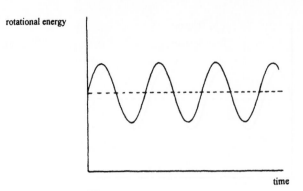

At the lower frequencies, rotational energy transfer keeps
the rotational energy changes in phase with the sound
wave.

**Figure 8.3a Collisions are occurring fast enough, and are effective in transferring rotational
energy to maintain a Maxwell-Boltzmann distribution.**

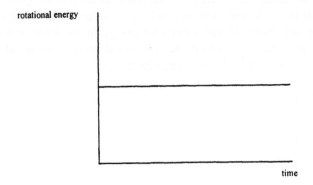

At higher frequencies, rotational energy transfer cannot
keep in phase with the sound wave and no transfer
occurs, leaving the rotational energy constant throughout
the disturbance created by the sound wave.

Figure 8.3b Collisions are not effective in transferring rotational energy.

The frequencies at which these situations occur should give a measure of the
effectiveness of collisions in transferring vibrational and rotational energy. These
frequencies are determined by studying the change in the velocity of sound, or the
change in the absorption of sound as the frequency of the sound wave through the gas,

or solution, is increased. A point of inflection in the graph of (velocity)^2 against frequency, or a maximum in the graph of absorption against frequency is found at the frequencies at which collisions become ineffective in transferring vibrational and rotational energy, Figures 8.4 and 8.5.

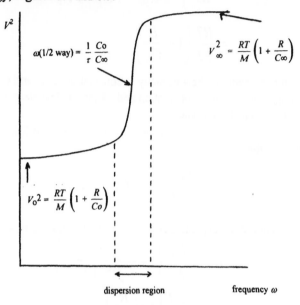

$$\alpha(1/2 \text{ way}) = \frac{1}{\tau}\frac{Co}{C\infty}$$

$$V_{\infty}^2 = \frac{RT}{M}\left(1 + \frac{R}{C\infty}\right)$$

$$V_0^2 = \frac{RT}{M}\left(1 + \frac{R}{Co}\right)$$

dispersion region   frequency ω

Figure 8.4   Dependence of velocity of sound on the frequency of a sound wave - one relaxation time at which collisions become ineffective in transferring vibrational or rotational energy.

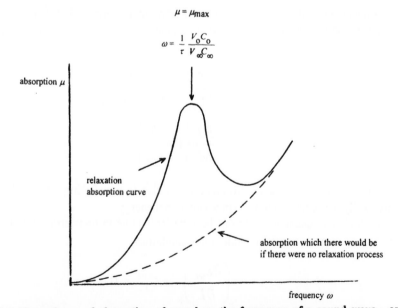

$$\mu = \mu_{max}$$

$$\omega = \frac{1}{\tau}\frac{V_0 C_0}{V_\infty C_\infty}$$

absorption μ

relaxation absorption curve

absorption which there would be if there were no relaxation process

frequency ω

Figure 8.5   Dependence of absorption of sound on the frequency of a sound wave - one relaxation time at which collisions become ineffective in transferring vibrational or rotational energy.

The velocity of sound through an ideal gas is related to the heat capacities for the gas.

$$V^2 = \frac{RT}{M}\left\{1 + \frac{R}{C_V^{\text{trans}} + C_V^{\text{rot}} + C_V^{\text{vib}}}\right\} \tag{8.7}$$

$$= \frac{RT}{M}\left(1 + \frac{R}{C_0}\right) \tag{8.8}$$

where C_V^{trans} is the translational heat capacity at constant volume, and C_V^{rot} and C_V^{vib} are the corresponding heat capacities for rotation and vibration. The electronic heat capacity is zero for the ground electronic state.

If the energy of vibration ε_{vib} is constant, which it is when collisions fail to transfer energy fast enough, then

$$\left(\frac{\partial \varepsilon_{\text{vib}}}{\partial T}\right)_V = C_V^{\text{vib}} = 0 \tag{8.9}$$

and so C_V^{vib} drops out of the expression (8.7) above and V^2 takes a new and higher value

$$V^2 = \frac{RT}{M}\left\{1 + \frac{R}{C_V^{\text{trans}} + C_V^{\text{rot}}}\right\} \tag{8.10}$$

$$= \frac{RT}{M}\left(1 + \frac{R}{C_\infty}\right) \tag{8.11}$$

$$\text{where } C_\infty = C_V^{\text{trans}} + C_V^{\text{rov}}. \tag{8.12}$$

This behaviour is illustrated for transfer of vibrational energy in Figure 8.4.

Corresponding behaviour is found at frequencies at which rotational energy transfers can be studied. Here

$$V^2 = \frac{RT}{M}\left\{1 + \frac{R}{C_V^{\text{trans}} + C_V^{\text{rot}}}\right\} \tag{8.13}$$

$$= \frac{RT}{M}\left\{1 + \frac{R}{C_0'}\right\} \tag{8.14}$$

describes behaviour when collisions are effective in transferring rotational energy, but are totally ineffective in transferring vibrational energy.

At higher frequencies collisions become ineffective in transferring rotational energy and so ε_{rot} remains constant. Under these conditions

$$\left(\frac{\partial \varepsilon_{\text{rot}}}{\partial T}\right)_V = C_V^{\text{rot}} = 0 \tag{8.15}$$

and so C_V^{rot} drops out of Equation 8.13 and V^2 takes a new and higher value.

$$V^2 = \frac{RT}{M}\left\{1 + \frac{R}{C_V^{\text{trans}}}\right\} \quad (8.16)$$

$$= \frac{RT}{M}\left\{1 + \frac{R}{C'_\infty}\right\} \quad (8.17)$$

Similar type equations will hold for solutions.

8.3.1 Determination of the relaxation time

In Figure 8.4 the graph shows a point of inflection at a particular frequency which lies between the low steady value and the high steady value of the square of the velocity of sound. The region between these two steady values is called the dispersion region, and it is in this region that the most useful information is found.

At the point of inflection at a frequency ω

$$\omega^2\tau^2 = \frac{C_0^2}{C_\infty^2}\left[\frac{V_\omega^2 - V_0^2}{V_\infty^2 - V_0^2}\right] \quad (8.18)$$

where τ is the relaxation time and $V_\omega^2, V_0^2, V_\infty^2$ are the squares of the velocity of sound at the point of inflection, the low frequency end and the high frequency end respectively.

Points of inflection are difficult to pinpoint accurately, and often the frequency exactly half-way between the low and high steady values is used.

$$\omega_{(\frac{1}{2}\text{way})} = \frac{1}{\tau}\frac{C_0}{C_\infty}. \quad (8.19)$$

All quantities, other than τ are known in Equations 8.18 and 8.19, and so τ can be found. Depending on the frequency region being studied the relaxation times for vibrational and rotational energy transfer can be determined.

If, instead of the velocity of sound, the absorption of sound is measured then a peak in the absorption is found, Figure 8.5. This maximum corresponds to the limit of effectiveness of collisions transferring vibrational energy or rotational energy. Vibrational energy drop out occurs at low frequencies with rotational drop out occurring at higher frequencies.

At the maximum

$$\omega_{\text{max}} = \frac{1}{\tau}\frac{V_0 C_0}{V_\infty C_\infty}. \quad (8.20)$$

Hence the relaxation time can again be found.

Since a maximum is more accurately determined than a point of inflection or a half-way point, absorption studies are often used in preference to velocity of sound studies.

Figures 8.6 and 8.7 give overall diagrams showing both vibrational and rotational fall off, using velocity of sound and absorption data respectively.

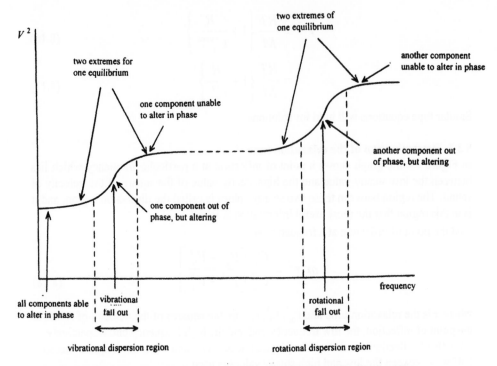

Figure 8.6 Velocity of sound diagram for vibrational and rotational relaxation.

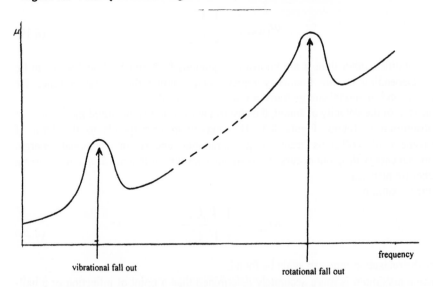

Figure 8.7 Absorption of sound diagram for vibrational and rotational relaxation.

8.3.2 Determination of rate constants for vibrational energy transfers studied by ultrasonic studies

If energy transfer between **all** levels is considered then the following equation can be derived.

$$\frac{1}{\tau} = f_{10} - f_{01} \tag{8.21}$$

where f_{10} and f_{01} are transition probabilities for vibrational energy transfer between the zero'th and first vibrational levels.

The corresponding rate constants for a system consisting of a total number, N, of molecules present, are

$$k_{01} = \frac{f_{01}}{N} \tag{8.22}$$

$$k_{10} = \frac{f_{10}}{N}. \tag{8.23}$$

The relaxation time involves the zero'th and first vibrational levels only, and ultrasonic experiments can only give information on transfer of energy between these levels, and there is no way in which the experiments can be altered to give data on transfer between higher levels. The restriction is a theoretical one, and not an experimental one.

A kinetic analysis involving energy transfer between all levels reduces to

$$\frac{f_{01}}{f_{10}} = \left(\frac{N_1}{N_0}\right)_{eq} \tag{8.24}$$

where N_1 is the number of molecules in the first excited level, N_0 is the number of molecules in the ground vibrational level, and "eq" signifies that these correspond to a Maxwell-Boltzmann distribution of vibration in these two levels where

$$N_1 = N_0 \exp\left(-\frac{h\nu}{kT}\right) \tag{8.25}$$

where ν is the frequency for the transition from the ground level to the first excited level.

From these equations

$$\frac{f_{01}}{f_{10}} = \exp\left(-\frac{h\nu}{kT}\right) \tag{8.26}$$

which taken with Equation 8.21 gives

$$\frac{1}{f_{10}} = \tau\left[1 - \exp\left(-\frac{h\nu}{kT}\right)\right] \tag{8.27}$$

and

$$\frac{1}{f_{01}} = \tau\left[\exp\left(+\frac{h\nu}{kT}\right) - 1\right] \tag{8.28}$$

from which f_{10} and f_{01} can be found from the experimental values of τ and ν.

Similar equations are used to treat rotational energy transfers.

Values for transition probabilities for vibrational energy transfer between higher levels, and where ascent of the vibrational energy ladder is by ladder climbing, can be estimated from Equation 8.5. In principle, all energy transfer transition probabilities and rate constants for ladder climbing processes can be found. However, the value for energy transfer rate constants between higher levels are limited in accuracy because the theoretical relation, Equation 8.5, assumes harmonic oscillators.

The transition probabilities and rate constants for rotational energy transfer between the zero'th and first rotational levels can likewise be found from the rotational relaxation time, and extension to higher levels is made using Equation 8.6 which covers both ladder climbing and strong collisions.

Two other useful quantities are

(i) the transition probability *per collision*, p_{10}, p_{01}

$$p_{10} = \frac{f_{10}}{Z}, \qquad p_{01} = \frac{f_{01}}{Z} \tag{8.29}$$

where Z is the number of collisions per unit time which a molecule undergoes, and

(ii) the mean number of collisions required before energy transfer occurs, Z_{10}, Z_{01}

$$Z_{10} = \frac{1}{p_{10}}, \qquad Z_{01} = \frac{1}{p_{01}}. \tag{8.30}$$

In vibrational energy transfer, the energy is transferred *from* a low lying mode in the zero'th level in activation, and *to* a low lying mode in the zero'th level in deactivation. Thereafter it is very rapidly transferred to other modes via an intermolecular mechanism. If two points of inflection or two peaks occur close together, then energy is independently entering two low lying modes.

With rotational energy transfer the energy is very rapidly redistributed between molecules.

8.4 Shock Tubes

This technique can be used to study fast reactions, but it can also be used to study energy transfer. The essential features are passage of a shock wave at supersonic speeds through the gas causing heating of the gas to a very high temperature, often around 10^3 - 10^4 K. The temperature rise occurs in around 10^{-6} s, and so must result from a few collisions only. The high temperature stays virtually constant for up to several milliseconds, and the observations must be made during this period.

The molecules present at these high temperatures are greatly excited, and the experiment follows the build up of the populations of these highly excited levels required by the new Maxwell-Boltzmann distribution for the final steady high temperature. This build up occurs at the expense of the populations of the lower state.

The steady high temperature is attained in 10^{-6} s, and translation to translation and translation to rotation energy transfer can easily occur within this time. But translation to vibration exchange is difficult and will not occur in 10^{-6} s, and so what is observed in the shock tube is the setting up of the high temperature Maxwell-Boltzmann populations in high vibrational levels, mainly by translation to vibration transfers.

This build up can be followed by absorption spectroscopy from which transition probabilities and rate constants can be found. Either the build up of the populations of various excited levels, or the decay of a low lying vibrational level can be followed. If the monitored level is being built up, then an absorption peak for this level will also be built up with time. If the monitored level is one from which molecules are being excited then an absorption peak for this level will decay with time. For both cases, a continual recording of the population of the vibrational level under study can be made since the population of the level determines the intensity of the monitoring absorption.

If the ground vibrational level is monitored by absorption, the population and hence the intensity of the monitoring absorption peak will reflect movement of molecules *out of* the ground state *to all* possible excited states, and of all molecules *into* the ground state *from all* possible excited states as dictated by the high temperature vibrational energy distribution. Similarly, if build up of an excited level is being monitored by absorption, then the intensity of absorption will reflect the number of molecules moving *into* and *out* of that state *from all* possible levels *below* and *above* the vibrational level which is being studied. In the experiment all observable levels can be monitored in this way.

From this the rate constant for build up of an excited level during the period of constant high temperature can be found. But, since only the *overall* population of each possible state is built up from *all* levels *below* and *above* the particular state, then only *overall* rate constants emerge. Ideally, the overall rate constant should be split up into detailed individual rate constants for each possible energy transfer contributing to the build up of any given excited level. This is not possible (Section 8.12), and can be contrasted with molecular beam experiments which can furnish individual state to state rate constants.

8.5 Flash Photolysis

Again this technique can also be used for studying fast reactions and for energy transfer.

An intense flash of light excites molecules out of the normal Maxwell-Boltzmann distribution into excited vibrational levels of the ground state, or into excited vibrational levels of excited electronic states. The molecules then lose their energy by a collisional mechanism of energy transfer, generally to translation, but sometimes to rotation, thereby returning to the original Maxwell-Boltzmann distribution.

In flash photolysis the energy of the flash is so large that the populations of the excited states are so high that conventional spectroscopic methods can detect and measure these levels. It then becomes possible to watch movement *out of* and *into* these levels by following the decrease of intensity of suitable peaks in absorption, as described in Section 8.4. Again, however, only an overall population can be measured, and so only overall rate constants can result.

Although shock tubes and flash photolysis both investigate the same fundamental situation, processes in the shock tube must be carefully distinguished from those in flash photolysis. In shock tube experiments a process of *excitation* to a new Maxwell-Boltzmann distribution corresponding to the high temperature is being studied, but in flash photolysis experiments the process is now one of *deactivation* back to the original Maxwell-Boltzmann distribution corresponding to the laboratory temperature.

Both shock waves and flash photolysis are indiscriminate in so far as a large range of levels are populated. In contrast, laser methods and chemiluminescence are more discriminate with only a few levels populated (Sections 8.7, 8.8).

8.6 Fluorescence
Here the molecule is excited into highly excited vibrational levels of the ground electronic state, or into a highly excited electronic level. Fluorescence then occurs to the ground level via various excited vibrational levels. Decay to these levels is very rapid, and in consequence sometimes cannot be observed. Experiments suggest that the levels populated in these ways often lie between $v = 0$ and $v = 5$. Vibrational relaxation to the ground vibrational level occurs by collisions, and this relaxation can be monitored by absorption.

In practice the rapid decay of the upper levels by fluorescence means that what is observed is the collisional relaxation between low lying levels, and this often reduces to the rate constant for deactivation from $v = 1$ to $v = 0$. However, improvements in the speed of detection will remove this limitation.

8.7 Laser Photolysis
Because of its high energy, a laser can produce populations in excited states which are grossly disturbed from their normal Maxwell-Boltzmann distribution. Furthermore, it excites into specific states which are known unambiguously. Pulsed lasers allow absorption of a number of photons successively so that the molecule can pass stepwise by a ladder climbing process into higher levels. Similar techniques using masers enable rotational states to be studied.

Once excited the molecule can fluoresce, and the intensity of fluorescence monitors the population of the excited level. Alternatively, the level could be monitored by absorption spectroscopy. Overall rate constants describing overall movement *into* and *out of* the level can be found. By choosing appropriate lasers and masers a whole variety of levels could be studied, though in practice this is limited by the availability of suitable lasers and masers. The energy transfers for rotations often cannot be observed independently because the Maxwell-Boltzmann distribution for rotations is set up so rapidly.

8.8 Chemiluminescence
Chemiluminescence is emission of radiation from the products of a chemical reaction, and is normally associated with exothermic reactions where the products are in excited vibrational and rotational levels, and sometimes in excited electronic states. Chemiluminescence helps to pinpoint the particular excited states which are produced *immediately* after reaction has occurred.

Unfortunately, collisional deactivation often has occurred before the initial states of the products can be detected. This is so for rotational states because of the rapidity of rotational energy transfer, resulting in the populations of the rotational levels being distorted from the levels produced immediately after reaction. Vibrational relaxation is not so serious a limitation, though it has been detected in some reactions. Using a flow system with several observation points gives a record of progressive relaxation which can be extrapolated back to give the populations at zero time after reaction.

Chemiluminescence is limited in that it can only be applied to reactions emitting light, but when applicable it is specific, though as with previously discussed methods it can only generate overall rate constants.

8.9 Steady state methods

This is an older method of obtaining energy transfer data, but problems in producing sufficiently high populations in excited levels, coupled with difficulties in analysis of the data limited its scope. With laser developments the method has been revived.

A vibrationally excited molecule M* can be produced by absorption of radiation or collisional activation. It can then be deactivated by collision, or can relax by emission of radiation.

$$M + h\nu \quad \underset{k_{fluorescence}}{\overset{k_{absorption}}{\rightleftharpoons}} \quad M^*$$

$$M^* + X \quad \underset{k_{activation}}{\overset{k_{deactivation}}{\rightleftharpoons}} \quad M + X$$

where M and M* is the molecule in its ground state and excited state respectively, and X is a third body.

A steady state treatment on M* gives

$$[M^*] = \frac{k_{absorp} I[M] + k_{act}[M][X]}{k_{fluor} + k_{deact}[X]} \tag{8.31}$$

where I is the intensity of the illumination

The removal of M* can be followed by measuring the total fluorescent intensity, F

$$F = k_{fluor}[M^*]$$

$$= \frac{k_{fluor}\{k_{absorp} I[M] + k_{act}[M][X]\}}{k_{fluor} + k_{deact}[X]} \tag{8.32}$$

Deactivation by collision increases as the concentration of third body increases, and so the total fluorescent emission is measured as a function of total pressure, enabling the rate constants to be found by successive approximations.

Approximation 1

If [X] is low and I is large, a first approximate solution to Equation 8.32 gives

$$\frac{1}{F} = \frac{k_{fluor} + k_{deact}[X]}{k_{fluor} k_{absorp} I[M]} \tag{8.33}$$

and

$$\frac{I[M]}{F} = \frac{1}{k_{absorp}} + \frac{k_{deact}}{k_{fluor} k_{absorp}}[X] \tag{8.34}$$

from which k_{absorp} can be found from the intercept, and $k_{deact}/k_{fluor} \, k_{absorp}$ from the gradient of a plot of $I\,[M]/F$ against $[X]$.

If k_{fluor} can be found from other experiments then first approximate values of all three rate constants can be found. These first approximate values substituted into Equation 8.32 give a first approximate k_{act}.

Successive reiteration will lead to more accurate values of the four rate constants defined in the two mechanistic steps.

If conventional sources are used, then an indiscriminate group of excited molecules are formed, and the rate constant describes the overall rate of energy transfer from all levels. With a precise laser one vibrationally excited level will be populated in contrast to the indiscriminate number of levels populated by conventional sources. And so deactivation of one level at a time can be followed by laser techniques. However, the rate constants are still overall rates in the sense discussed in Sections 8.2 and 8.4, and genuine state to state rate constants cannot be found.

8.10 Results from all methods other than molecular beams

Results indicate that transfer of vibrational energy between excited vibrational levels tends to be fast, with a bottleneck at $v = 1$ when there is a slow rate-determining step for transfer between the first and ground vibrational levels. This means that often the overall rate constant for collisional vibrational-vibrational and vibrational-translational energy transfer involving excited vibrational levels approximates to that for the rate-determining step for vibrational-translational energy exchange between $v = 1$ and $v = 0$, and so rate constants for the higher levels cannot be picked up. Increase in speed of detection should improve this situation. These transfers in the higher levels are thought to occur by resonant or near resonant vibrational-vibrational energy transfer, though there is evidence for resonant or near resonant vibrational-rotational energy transfer.

8.11 Comparison of Methods

Shock tubes and flash photolysis are indiscriminate in so far as a large number of excited states are populated, and these have to be found by analysis. In contrast, laser production of the excited state is highly selective, and detection of the excited state is straightforward using a secondary laser flash which can occur so fast that no relaxation takes place. Hence overall rate constants for all levels can be obtained in laser experiments, and in addition it is often possible to specify which particular vibrational modes are being excited, something which is not possible with shock tubes and flash photolysis.

Data from shock tubes, flash and laser photolytic apparatus sometimes only give the rate constant for the rate determining step of energy transfer between the zero'th and first vibrational levels, but more sophisticated, higher response fast detection equipment will allow the fast vibrational energy transfers in the higher levels to be studied. Ultrasonic methods can only generate the rate constant for energy transfer between the ground and first excited state.

All these methods give overall rate constants for energy transfer into and from a given level. Molecular beams, on the other hand, does give information on energy transfer between specified energy levels, and is the only technique which can give state to state rate constants.

8.12 Network arrays of rate constants in energy transfers

Methods which can only give overall rate constants look at energy transfer *to* and *from* one particular excited state, and do so by following movement of molecules *into* the excited state *from* all possible other states, and movement of excited molecules *out of* the excited state *into* all possible other states. This is in contrast to state to state rate constants which describe energy transfer between two specific states only. As discussed previously, molecular beams is the only technique which can give state to state rate constants.

State to state kinetics also cannot be inferred from the experimental data of the techniques given in Sections 8.4 to 8.9. In these methods the changing concentration of the excited molecule is monitored, and a whole series of differential equations can be generated describing formation of the excited state from all possible states and removal to all possible states

$$\frac{d[v,J]}{dt} = \left\{ \begin{array}{c} \text{rates of all steps} \\ \text{leading to } v,J \end{array} \right\} - \left\{ \begin{array}{c} \text{rates of all steps} \\ \text{leading from } v,J \end{array} \right\} \qquad (8.35)$$

where v and J denote vibrational and rotational states respectively.

In these equations there are more unknowns that there are equations, and so these equations cannot be solved to give the individual state to state rate constants. What is needed is not more experimental data, but a theory which would calculate enough of the rate constants explicitly, or give enough relations between the rate constants so that there will be enough equations to solve for all unknowns. At present, theory has not advanced sufficiently for this to be possible.

This problem does not arise in determination of rate constants from conventional kinetics experiments where either explicit integrated rate expressions, differential rate expressions or steady state treatments often suffice. There is a fundamental difference between such situations and state to state studies. In the conventional studies the rate constants are linked in a linear fashion. For example

$$\begin{array}{ccccccc} & k_1 & & k_2 & & k_3 & \\ A & \rightarrow & B & \rightarrow & C & \rightarrow & D \end{array}$$

where the steps are in a chain, and there are three constants to be found and only three independent concentrations which need be monitored. Hence solution can be achieved.

State to state mechanisms give a network of rate constants in contrast to the chain arrangement illustrated above. In the network there is a large difference between the number of rate constants involved, and the number of independent concentrations to be monitored. A considerably simplified mechanism

$$\begin{array}{ccccccc} A_1 & \rightarrow & A_2 & \rightarrow & A_3 & \rightarrow & A_4 & \cdots J = 0 \\ \downarrow & & \downarrow & & \downarrow & & \downarrow & \\ A_1' & \rightarrow & A_2' & \rightarrow & A_3' & \rightarrow & A_4' & \cdots J = 1 \\ \downarrow & & \downarrow & & \downarrow & & \downarrow & \\ A_1'' & \rightarrow & A_2'' & \rightarrow & A_3'' & \rightarrow & A_4'' & \cdots J = 2 \\ \vdots & & \vdots & & \vdots & & \vdots & \\ v = 0 & & v = 1 & & v = 2 & & v = 3 & \end{array}$$

where A_1, A_2, A_3 and A_4 represent molecule A in different vibrational states $v = 0$, $v = 1$, $v = 2$ and $v = 3$ respectively. A_1, A_1', A_1'' represent molecule A in different rotational states $J = 0$, $J = 1$ and $J = 2$ respectively. This network involves 17 rate constants and 12 populations, and the equations cannot be solved even if all 12 populations could be monitored independently. Only activation steps are given, for example $A_1' \rightarrow A_1''$ represents rotational activation. The corresponding deactivation, $A_1'' \rightarrow A_1'$, is not given in the network. But these are linked by the Maxwell-Boltzmann expression

$$\frac{k_{J=1 \rightarrow J=2}}{k_{J=2 \rightarrow J=1}} = \exp\left(-\frac{hv}{kT}\right)$$

(8.36)

where v is the frequency for the rotational transition between A_1' and A_1'' which can be measured spectroscopically.

What is more severely limiting is that the network does not include energy transfer such as $A_2 \rightarrow A_4''$ which involves vibrational activation from $v = 1$ to $v = 3$ along with simultaneous rotational activation from $J = 0$ to $J = 2$, or $A_1'' \rightarrow A_4$ which represents vibrational activation between $v = 0$ and $v = 3$ but simultaneous rotational deactivation from $J = 2$ to $J = 0$. Inclusion of energy transfers such as these would mean an even greater number of rate constants, but with the same number of populations to be monitored.

8.13 An outline of theoretical calculations on state to state rate constants, and use of these to formulate a theoretical expression for the overall observed rate of reaction

The general mechanism is

$$A + A \quad \xrightarrow{k_1} \quad A^* + A \qquad \text{activation}$$

$$A^* + A \quad \xrightarrow{k_{-1}} \quad A + A \qquad \text{deactivation}$$

$$A^* + bA \quad \xrightarrow{k_2} \quad \text{products} \qquad \text{reaction}$$

There is a whole series of energy levels

$$0, 1, 2, 3 \ldots \qquad n - 1, n, n + 1 \ldots$$

where n is the last level below the critical energy
          $n + 1$ is the first level above the critical energy

and r denotes any level below the critical, $0 < r < n$
          s denotes any level above the critical $s > n + 1$
with associated energies

$$\varepsilon_0, \varepsilon_1, \varepsilon_2, \varepsilon_3 \ldots \varepsilon_{n-1}, \varepsilon_n, \varepsilon_{n+1} \ldots$$

and populations per unit volume

$$n_0, n_1, n_2, n_3 \ldots n_{n-1}, n_n, n_{n+1} \ldots$$

Activation and deactivation rate constants for energy transfer between these levels can be displayed as on page 231

(i) Calculation of the rate of activation
The number of molecules in level s formed per unit time per unit volume from molecules in level r is

$$k_{r,s} n_r N$$

where N is the total number of molecules present per unit volume
   The total number of molecules with energy $\geq$ the critical energy formed per unit time per unit volume from the molecules in the various levels below the critical energy is

$$\sum_r \sum_s k_{r,s} n_r N$$

where the sum is over all possible levels above the critical, $s \geq n + 1$, and this is summed over all possible levels below the critical energy $0 \leq r \leq n$. These two summations mean that all possible instances of activation are covered.

$$\text{The total rate of activation} = \sum_{0 \leq r \leq n} \sum_{s \geq n+1} k_{r,s} n_r N. \tag{8.37}$$

(ii) Calculation of the rate of deactivation
The number of molecules in level r formed per unit time per unit volume from level s is

$$k_{s,r} n_s N.$$

The total number of molecules with energy $\leq$ the critical energy formed per unit time per unit volume from the molecules in the various levels above the critical energy is

$$\sum_s \sum_r k_{s,r} n_s N$$

where the sum is over all levels below the critical, $0 \leq r \leq n$, and this is summed over all possible levels above the critical, $s \geq n + 1$. The two summations mean that all possible instances of deactivation are covered.

$$\text{The total rate of deactivation} = \sum_{s \geq n+1} \sum_{0 \leq r \leq n} k_{s,r} n_s N. \tag{8.38}$$

(iii) **Calculation of the rate of reaction**

$$\text{The total rate of reaction } = \sum_{s \geq n+1} k_s n_s N^b$$

(8.39)

where k_s is the rate constant for reaction of an activated molecule which lies in level s where $s \geq n + 1$, n_s is the number of activated molecules per unit volume in level s, and N^b appears in the expression since k_s is a $(b + 1)$th order rate constant describing the $(b + 1)$ molecular reaction step, for example if $b = 0$, then $b + 1 = 1$ and reaction is unimolecular.

8.13.1 Carrying out these calculations: activation and deactivation
(a) The rate of activation
To calculate this rate all conceivable k_{rs} and n_r are required. The rate constants are found from tabulated transition probabilities, f_{rs}, where

$$k_{rs} = \frac{f_{rs}}{N}.$$

(8.40)

As shown previously molecular beam experiments have the greatest potential for generating the required transition probabilities.

The values n_r come from solution of the following equation

$$\frac{dn_r}{dt} = \sum_{q \neq r} k_{q,r} n_q N - \sum_{q \neq r} k_{r,q} n_r N - \sum_{s \geq n+1} k_{r,s} n_r N + \sum_{s \geq n+1} k_{s,r} n_s N$$

(8.41)

where q and r are any two different levels below the critical.

These four terms have the physical meaning given schematically in Figure 8.8, and verbally in the following statements.

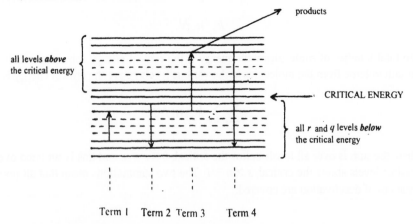

Figure 8.8   Schematic representation of the algebraic terms appearing in a calculation of the rate of activation.

(i)   The *first* term represents the rate of formation of molecules in the r'th level from all other levels below the critical which are greater or less than r.

(ii) The *second* term represents the rate of conversion of molecules in the r'th level to all other levels below the critical which are greater or less than r.

(iii) The *third* term represents the rate ⌐f conversion of molecules in the r'th level to all levels above the critical level, and thence to products.

(iv) The *fourth* term represents the rate of removal of all molecules from level s to molecules in level r below the critical energy.

There is a whole family of such equations, one for each r and each s level; each individual equation has to be summed over the appropriate sum to generate the complete family of equations which has to be solved to give the various required n_r either exactly by solution of the differential equation, or approximately using the steady state treatment.

(b) The rate of deactivation

Here all $k_{s,r}$ and n_s are required. As in activation all $k_{s,r}$ can, in principle, be found from molecular beam experiments, and all n_s from solution of the following equation

$$\frac{dn_s}{dt} = \sum_r k_{r,s} n_r N - \sum_r k_{s,r} n_s N + \sum_p k_{p,s} n_p N$$

$$- \sum_p k_{s,p} n_s N - k_s n_s N^b .$$

(8.42)

These five terms have the physical meaning given schematically in Figure 8.9, and verbally as follows

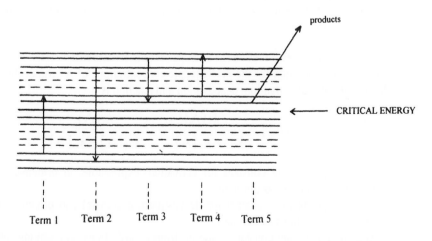

Figure 8.9   Schematic representation of the algebraic terms appearing in a calculation of the rate of deactivation.

(i)   The *first* term represents the rate of formation of all molecules in level s from all levels r below the critical level.

(ii)   The *second* term represents the rate of conversion of all molecules in level s to all levels r below the critical level.

(iii)   The *third* term represents the rate of formation of molecules in level s from all levels p which are above the critical, but are not s.

(iv)   The *fourth* term represents the rate of conversion of molecules in level s to all levels p which are above the critical, but are not s.

(v)   The *fifth* term represents the rate of removal of molecules in level s to products.

As in the rate of activation there is a whole family of such equations, where each individual equation is summed over appropriate r or s to generate the complete family of equations. These likewise have to be solved either explicitly, or by approximation using the steady state treatment.

8.13.2 Rate of the reaction step

Here all k_S and all n_S are required. The k_S values are rate constants for reaction from each s level. These rate constants can either be found experimentally by the methods described in this chapter and Chapter 9, or can be calculated theoretically from transition state theory, Chapters 2 and 3, or from unimolecular theory, Chapters 4 and 5.

The n_S values must come from solution of the equations given in the previous section for the rate of deactivation.

8.13.3 Generalisation of these calculations

In principle this is a mathematical scheme for calculating the rate of activation, deactivation and reaction for any simple elementary reaction where the only essential difference between the possible reactions is in the value of b, which defines the molecularity of the reaction step. These theoretical expressions can, in principle, cover all steps in even the most complex mechanism. What is required to calculate these rates are values of transition probabilities for all possible energy transfer steps, populations of all levels involved, and values of the rate constants describing reaction from all possible levels above the critical.

8.13.4 Relation of this treatment to macroscopic treatments

The general expression for the macroscopic concentrations of A^* molecules in terms of the overall rates of activation, deactivation and reaction is

$$[A^*] = \frac{k_1[A]}{k_{-1} + k_2[A]^{b-1}} = \frac{k_1[A]^2}{k_{-1}[A] + k_2[A]^b}. \tag{8.43}$$

k_1 is a composite rate constant for activation, expressible, in principle, in terms of transition probabilities as outlined.

k_{-1} is a composite rate constant for deactivation, expressible, in principle, in terms of transition probabilities as outlined.

k_2 is a composite rate constant made up from contributions representing rate constants from all levels above the critical.

From this, the overall rate is given by

$$k_2[A^*][A]^b. \tag{8.44}$$

9

Energy Transfer

Energy transfer is intimately linked with activation and deactivation processes, and, in consequence, has become an integral part of kinetics.

Methods of studying energy transfer have been discussed in Chapter 8 where it was shown that in most of the methods only overall rate constants could be found for transfer from specific quantum states, and that if detection was not sufficiently rapid only the rate-determining step of transfer between the zero'th and first vibrational levels could be picked up. Ultrasonics is even more limited, and in principle, can only quantify this rate determining step. In contrast molecular beams experiments are able to look at state to state transfers individually.

This chapter will limit discussion to vibrational and rotational energy transfer. Transfer of electronic energy will not be discussed even though it is highly pertinent to photochemical reactions.

Collisions transfer energy, and efficient vibrational energy transfer occurs when:
    the duration of a collision < the time for one vibration,
and efficient rotational energy transfer occurs when:
    the duration of a collision < the time for one rotation.

The time for vibration is long when the frequency of vibration is low, and this favours efficient transfer. The wide range of fundamental vibration frequencies found for many substances suggests that there will be a wide range in transfer efficiency.

The time for rotation is long for a slowly rotating molecule, and is a consequence of a large moment of inertia. Efficient rotational energy transfer occurs for molecules with high moments of inertia and small spacings. Most polyatomic molecules fit this criterion, and transfer occurs on nearly every collision. H_2 and hydrides have low moments of inertia, and for them rotational relaxation is slower.

The efficiency of transfer of vibrational energy increases as the energy of vibration increases and as the vibrational quantum number increases, and so the higher levels relax faster than the lower levels. The amplitude of vibration also increases with increasing energy, and so the more vigorously the molecule vibrates the more efficient transfer will be. High temperatures favour efficient transfer partly because of the cooperative effect of increased energy, increased amplitude, and a decrease in the duration of a collision.

The rotational frequency increases rapidly with increasing energy and so the time for one rotation gets less, resulting in less efficient rotational energy transfer. Relaxation, therefore, becomes more difficult as the rotational energy ladder is climbed, in contrast to the increased relaxation for higher values of the vibrational quantum number. As the temperature increases, the duration of a collision decreases, and rotational relaxation becomes easier at high temperatures.

These arguments apply to the transfer of rotational energy from one specific quantum state to another specific quantum state, and so are only relevant to molecular beam studies. But for the other techniques from which only an overall transition probability is found, the opposite is true. Because the population of the more slowly relaxing higher levels increases with increasing temperature, the effects of these higher levels dominate at high temperatures giving an overall decreased rate of relaxation in contrast to the increased rate at high temperatures expected for the state to state transfers.

Table 9.1 shows results for transfer of different types of energy, where Z is the number of collisions required to de-excite one quantum of energy. These are only "typical" values, and there is a wide spread in the values for each category of transfer.

Table 9.1 Typical values for energy transfers in gases

transfer between	relaxation time at 1 atm pressure	Z
rotation and rotation	10^{-10}s	<10
vibration and vibration	10^{-8}s	100
rotation and translation	10^{-8}s	100
vibration and rotation	10^{-6}s	10^4
vibration and translation	10^{-4}s	10^6

Tables 9.2 and 9.3 give values of Z_{10}, the number of collisions required to deactivate the first vibrational level to the ground state, for diatomic molecules and polyatomic molecules respectively. Energy transfer is a difficult process for a diatomic molecule, but becomes much easier for polyatomic molecules.

Table 9.2   Values of Z_{10} for diatomic molecules at 300K for vibration to translation transfer (ultrasonic, shock, flash and laser studies)

Z_{10}	1×10^{10}	1×10^8	$>2.7 \times 10^7$	2×10^7	1×10^7
gas	CO	D_2	N_2	O_2	H_2
Z_{10}	7×10^4	8×10^3	8×10^3	5×10^3	79
gas	Cl_2	HCl	HBr	Br_2	HF

Table 9.3 Values of Z_{10} for polyatomic molecules at 300K

Z_{10}	5.3×10^4	1.5×10^4	1.0×10^3	890	575
gas	CO_2	CH_4	SF_6	PH_3	CH_3Cl
Z_{10}	250	43	10	2.5	
gas	SO_2	C_6H_6	C_2H_6	$SnCl_4$	

Theory predicts that energy transfer should be faster, and therefore require fewer collisions, the smaller the amount of energy to be transferred. The rate determining step for vibrational relaxation is normally a vibration to translation transfer from $v = 1$ to $v = 0$ for diatomic molecules, and from the lowest lying mode for polyatomic molecules. Figure 9.1 shows this to be the case.

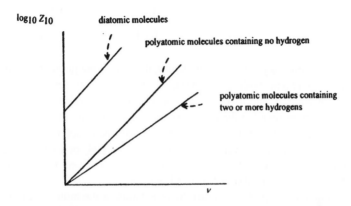

$\log_{10} Z_{10}$ diatomic molecules

polyatomic molecules containing no hydrogen

polyatomic molecules containing two or more hydrogens

v

Figure 9.1 Graph of log Z_{10} against v, showing that the smaller the energy gap, $h\nu$, the lower is the number of collisions required to deexcite one quantum of vibrational energy Z_{10}.

9.1 Theories of energy transfer

Early theories of vibrational energy transfer consider only the repulsive part of the intermolecular potential energy with attractions being neglected. An expression for P_{10}, the transition probability, Z_{10}, the number of collisions required to transfer a quantum of vibrational energy, and their temperature dependence follows, for energy transfer between the zero'th and the first level.

The theory predicts that

$$\log P_{10} \text{ v } T^{-1/3}$$

or $\qquad \log Z_{10} \text{ v } T^{-1/3}$

are linear, with deviations expected at temperatures at which attractions can no longer be ignored. These deviations occur at low temperatures for non-polar molecules, and extend to much higher temperatures for polar molecules where there are considerable attractive forces between the molecules. Experimental results are in agreement with theory, and suggest that attractive forces in polar molecules make transfer of vibrational energy much easier. The theory also predicts that vibration to translation energy transfer becomes faster as the vibrational ladder is climbed, with a slow rate-determining step between the zero'th and first levels resulting in one relaxation time only.

Even an elementary theory for rotational energy transfer runs into considerable difficulties, mainly because of the complicated motion involved, and because many rotational levels are occupied at room temperature. Theory predicts a single relaxation time, which unlike vibration, cannot be linked to any one transition, but involves all rotational levels. Table 9.4 gives some typical values of the number of collisions required to de-excite one quantum of rotational energy. The number of such collisions is also expected to increase with temperature.

Table 9.4 Values of Z_r for other gases at 300K

Z_r	100	30	15	5	5
gas	H_2	H_2S	CH_4	C_3H_6	NH_3
Z_r	4	2.5	1.5	1.3	
gas	H_2O	CO_2	C_2H_4	C_3H_8	

The Schwartz, Slawsky and Herzfeld theory, though giving good qualitative and semi-quantitative agreement with experiment, is inadequate in several respects. Again attractions are ignored, and there is also considerable uncertainty in the intermolecular potential energies assumed at the close distances of a collision. In addition, a spherically symmetrical intermolecular potential energy is assumed, but this is rarely the case except for spherically symmetrical molecules such as CH_4 or CCl_4. These difficulties can be removed if intermolecular potential energies derived from molecular beam experiments are used instead of the theoretical ones.

Another fundamental problem is that the orientation of the molecules is assumed to be constant during a collision. This is not valid for molecules with low moments of inertia which rotate rapidly and have wide spacings in the rotational levels, such as H_2 and hydrides. Rapid rotation enables the direction of approach to average out during a collision, consequently the orientation cannot be in a single direction.

Slow rotation, on the other hand, allows the direction of approach to be effectively constant. During the collision the molecule will not have rotated much, and hence can present a constant orientation. This applies to molecules with large moments of inertia and low spacings, and this category covers a large number of compounds, both diatomic and polyatomic.

The theory is therefore more adequate for diatomics and polyatomics which do not include H or D, and is inadequate for all molecules containing H or D.

One further serious problem is that the theory ignores the possibility of a simultaneous rotational energy transfer whenever collisional vibrational relaxation to translation occurs. All molecules have rotation, and it is likely that molecules change their rotational states on collision. Like the orientation assumption, this is particularly important for molecules with low moments of inertia and wide rotational spacings, such as found for H_2, D_2, hydrides and deuterides. It is expected that omission of rotational effects for these molecules would be serious, though it may be less important for molecules with large moments of inertia and small spacings.

Rotational energy transfers either facilitate vibrational-translational energy transfer, or do not. Transfer of energy to another molecule is easier the smaller the amount of energy transferred, and so, if simultaneous rotational transfer decreases the total amount of energy transferred to translation, then vibrational-translational energy transfer will be facilitated. If, on the other hand, simultaneous rotational transfer increases the total amount of energy transferred to translation, then vibrational-translational energy transfer is not helped.

The need to include the possibility of rotational energy transfers can be most easily demonstrated by diagrams, Figures 9.2 and 9.3. These are given for a diatomic molecule where the molecule loses energy to another molecule. These diagrams and the arguments which follow are given for the case of a harmonic oscillator. Anharmonicity effects should normally be considered

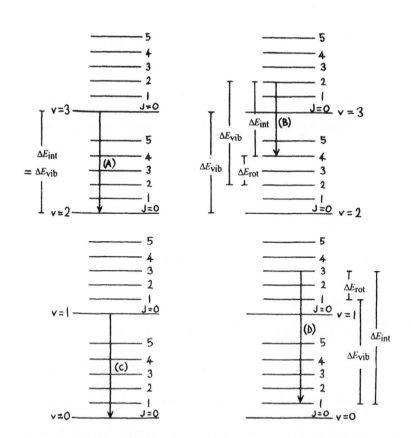

Figure 9.2 Possible vibrational changes for a diatomic molecule with low moment of inertia and wide spacing (a) with no rotational contributions (b) with rotational contributions.

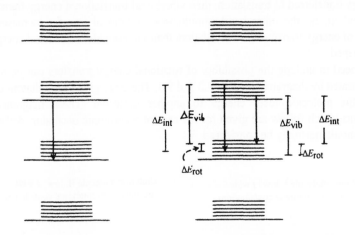

Diatomic molecule, high I, small
spacings: no rotational contributions

Diatomic molecule, high I , small
spacings: with rotational contributions   ions

Figure 9.3 Possible vibrational changes for a diatomic molecule with high moment of inertia and small spacings (a) with no rotational contributions (b) with rotational contributions.

9.1.2 Rotational contributions to vibration to translation energy transfer

(i) *Rotational energy transfer facilitates collisional vibration to translation transfer*

Transition A

1. $v = 3, J = 0 \rightarrow v = 2, J = 0$
2. Purely vibration to translation transfer
3. No rotational energy change
4. Energy transferred = ΔE_{vib}

Transition B

1. $v = 3, J = 2 \rightarrow v = 2, J = 4$
2. Both vibrational and rotational energies change
3. Energy transferred to translation = ΔE_{int}
4. ΔE_{int} = vertical distance between states $v = 3, J = 2$ and $v = 2, J = 4$ this is the *actual* transition involved
5. Energy change if the pure vibration to translation transfer were involved = ΔE_{vib}, see transition A.
6. ΔE_{vib} = vertical distance between $v = 3, J = 2$ and $v = 2, J = 2$ or between $v = 3, J = 0$ and $v = 2, J = 0$.
7. $\Delta E_{vib} > \Delta E_{int}$ by an amount equal to the vertical distance between $J = 4$ and $J = 2$ which is ΔE_{rot}

8. $\Delta E_{int} = \Delta E_{vib} - \Delta E_{rot}$ so the energy transferred to translation in deactivating $v = 3$ when simultaneous rotation is involved, is less than that when there is no rotational contribution.

9. Since $\Delta E_{int} < \Delta E_{vib}$ simultaneous rotational energy transfer facilitates vibrational-translational transfer.

10. This energy transfer, with simultaneous rotational energy transfer, will be *faster* than the pure vibrational-translational energy transfer, $v = 3 \rightarrow v = 2$, transition A.

11. In the *faster* transfer, v and J are changing in opposite directions, v is decreasing 3 $\rightarrow$ 2, J is increasing 2 $\rightarrow$ 4.

12. Vibrational transfer, with simultaneous rotational energy transfer is the preferred route, since it involves a smaller energy transfer to translation.

(ii) *Rotational energy transfer does not facilitate collisional vibration to translation transfer*

This situation is summarised in transitions C and D, and can be argued similarly to case (i) above.

Transition C represents energy transfer between $v = 1, J = 0$ and $v = 0, J = 0$, and so is a pure vibration to translation transfer.

Transition D represents simultaneous vibrational-rotational energy transfer, but now

$$\Delta E_{int} = \Delta E_{vib} + \Delta E_{rot}. \tag{9.1}$$

Since $\Delta E_{int} > \Delta E_{vib}$, simultaneous rotational energy transfer does not facilitate vibrational-translational energy transfer, and so transition D will be slower than the pure vibration to translation energy transfer $v = 1$ to $v = 0$, transition C. In this slower transfer, v and J are changing in the same direction, v is decreasing from $1 \rightarrow 0$ and J is decreasing from $3 \rightarrow 1$. The pure vibrational translational energy change will be the preferred route since it involves a smaller energy transferred to translation.

A similar analysis can be carried out for diatomic molecules with high moments of inertia and small spacings. Because of the small spacings ΔE_{rot} for a given change in J is much smaller than in a molecule with wide spacings. ΔE_{rot} is, therefore, a correspondingly smaller fraction of ΔE_{vib}, and causes a much smaller difference between ΔE_{vib} and ΔE_{int}. Hence effects from simultaneous rotational transfer have a less dramatic effect on the relaxation of a high moment of inertia compared to a low moment of inertia, molecule, Figure 9.3.

9.2 Extension of the theoretical basis to polyatomic molecules

Most polyatomic molecules are flexible and change shape during vibration and on collision. This is a serious problem, and is extremely difficult to handle theoretically. Furthermore, attractions and dipole effects in polar molecules cause problems, while the creation of transient dipoles during collisions of non-polar molecules is equally tricky to handle.

The complex motion of the atoms in a polyatomic molecule can be resolved into various normal modes of vibration. One immediate consequence of this is that it allows the possibility of vibrational-vibrational energy transfer which, in turn, provides an easier and faster route to the relaxation of vibrational energy in a polyatomic molecule, in direct contrast to the impossibility of vibration to vibrational transfer for the single vibrational mode of a diatomic molecule. The more normal modes there are the more efficient the vibrational relaxation can become, so instead of single modes relaxing independently with multiple relaxation times, one for each mode, the whole vibrational manifold can relax in unison by vibrational-vibrational energy transfer until the first excited state of the lowest mode is reached. At this stage vibrational-vibrational energy transfer is no longer possible within the molecule, and the final step is transfer of vibrational energy to translational energy in a slow rate-determining step. Relaxation using this sort of mechanism gives one relaxation time.

Laser experiments have confirmed the expectation from theory that for excited states vibrational to vibrational energy transfer is faster than vibration to translation transfer, and that both these types of transfer are much faster than vibration to translation transfer from the first excited state of the lowest mode.

9.3 Resonant and non-resonant transfers

In vibration to vibration transfers between two molecules, one of the molecules loses vibrational energy ΔE_1 while the second partner in the collision gains vibrational energy ΔE_2

(i)   If $\Delta E_1 = \Delta E_2$, then the change is resonant, and no excess vibrational energy is transferred to translation.

(ii) If $\Delta E_1 \neq \Delta E_2$, then the change is non resonant, and the excess vibrational energy is transferred to translation.

(iii) If $\Delta E_1 \approx \Delta E_2$, then the change is near resonant, and only a small amount of energy is transferred to translation.

(iv) If $\Delta E_1 > \Delta E_2$, then vibrational energy equal to $\Delta E_1 - \Delta E_2$ is transferred to translation, and is shared by the colliding molecules.

(v)   If $\Delta E_1 < \Delta E_2$, then translational energy equal to $\Delta E_2 - \Delta E_1$ has to transfer to vibration from the colliding molecules to allow the transfer to take place.

One feature of vibration to vibration transfer is that large amounts of energy can be transferred, but only if the transfer is resonant or near resonant, as only then is the amount of energy transferred to translation small. In a vibration to vibration transfer it is the size of the energy transfer to translation which determines the rate of transfer, not the amount of vibrational energy transferred between two modes. However, for effective and worthwhile deactivation the amount of vibrational energy exchanged should be large, and the transfer should be fast, with the latter criterion forcing the transfer to be resonant.

The bigger the amount of energy transferred to translation the smaller is the probability of the overall transfer, and a greater number of collisions is required to transfer one quantum of energy. A transfer which can minimise the amount of energy transferred to translation will increase the relaxation rate by providing an easier route to relaxation. One which increases the amount of energy transferred to translation decreases the relaxation rates and will not be utilised. Hence resonant or near resonant transfers are faster than non-resonant transfers. But within this group of fast resonant vibration to vibration transfers there is a spectrum of rates, with the fastest being those transferring the smallest amount of vibrational energy.

9.4 Vibration to vibration energy transfers

The greater efficiency of vibrational to vibrational energy transfer compared to vibrational energy transfer to translation can be most easily demonstrated on an energy diagram constructed from a consideration of the normal modes of vibration. In a molecule with three modes there are three vibrational frequencies v_1, v_2 and v_3 corresponding to modes 1, 2 and 3. Each of these lie at a given energy. Second harmonics of each lie at twice the energy of this state. Other quantum states are produced by combinations of the various contributing modes, Figure 9.4.

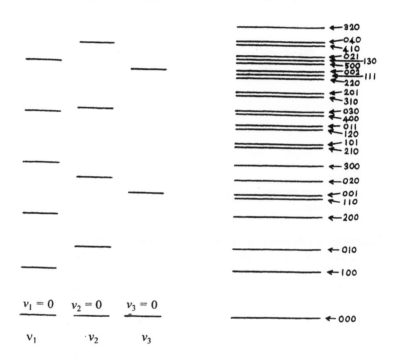

Figure 9.4   Energy diagram for a simple polyatomic molecule.

9.4.1 Vibrational energy transfer to translation during a collision

Transfer of vibrational energy to translation *without* any other vibration to vibration change can only occur if one mode changes quantum number while the other two remain constant. Transition A, Figure 9.5.1 fits this. Energy is removed from mode v_1 while the other two modes remain unchanged, and the overall change is $v_1 = 2$, $v_2 = 1$, $v_3 = 0 \rightarrow v_1 = 0$, $v_2 = 1$, $v_3 = 0$. The energy change is large with a consequent low probability, and the energy is transferred to translation in the other partner of the collision.

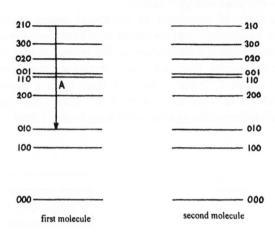

first molecule   second molecule

undergoes no change of rotational state

Figure 9.5.1 A simple vibration to translation energy transfer between two molecules on collision.

9.4.2 Vibration to vibration energy transfer during a collision

Figure 9.5.2 shows a resonant vibration to vibration change between two molecules 1 and 2 during collision.

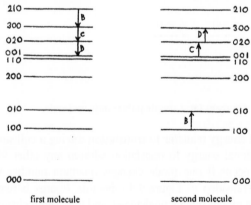

first molecule   second molecule

Figure 9.5.2 Resonant vibration to vibration energy transfer between two molecules on collision.

The molecule can climb down the energy ladder via a series of fast vibration to vibration steps where the energy is transferred in a resonant collision to another molecule. Transitions B, C and D in Figure 9.5.2 illustrate this. These represent small changes of energy, but the accumulative effect is to transfer a very large amount of energy in a series of fast resonant transfers which can very effectively compete with the large, but slow, transfers which accompany pure vibration to translation transfer.

In this type of transfer, more than one mode changes quantum number, in transition B the change in molecule 1 is 210 → 300 and in molecule 2 is 100 → 010. This means that in molecule 1 mode 1 gains energy and mode 2 loses energy, while in molecule 2 mode 1 loses energy and mode 2 gains energy. Similar considerations apply to transitions C and D.

Figure 9.5.3 shows several non-resonant vibration to vibration exchanges which will occur with simultaneous vibrational-translational energy changes. However, because these are non-resonant transfers the probability of transfer will be much less than that for the resonant transfers which are the favoured route for deactivation in molecule 1 and activation in molecule 2.

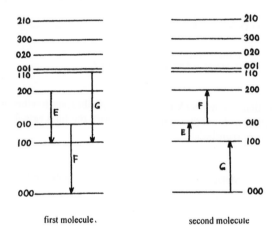

Figure 9.5.3   Non-resonant vibration to vibration energy transfer between two molecules on collision.

9.5 Intra and intermolecular vibration to vibration transfers in diatomic molecules

There is no route for *intra* molecular vibrational to vibrational energy exchanges on collision in a diatomic molecule because there is only one mode of vibration.

In contrast, intermolecular transfer of vibration to vibration on collision is possible in diatomic molecules, and can be a very effective means of transferring energy. Harmonic vibrations give very fast resonant energy transfers. Anharmonic vibrations change these resonant transfers into non-resonant transfers, while contributions from rotation can convert these non-resonant transfers into resonant transfers or vice-versa, Figures 9.6 and 9.7.

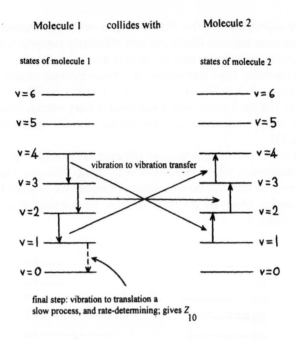

Figure 9.6 **Deactivation of a molecule (1) by resonant vibration to vibration energy transfers on collision with another molecule, (2) with the final rate-determining step being vibration to translation.**

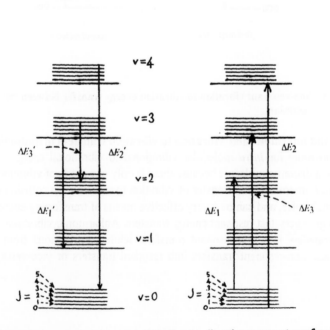

Figure 9.7 **Resonant and non-resonant vibration to vibration energy transfers in a diatomic molecule with rotational contributions.**

9.5.1 Resonant vibration to vibration energy transfers in diatomic molecules

Figure 9.6 shows resonant vibrational energy transfers between harmonic oscillators. These can be summarised as follows.

Changes in molecule 1 on collision with molecule 2

1. Changes quantum state from
   $v = 4$ to $v = 3$
2. Change in vibrational energy =
   $\Delta E_{4 \to 3}$

Changes in molecule 2 on collision with molecule 1

1. Changes quantum state from
   $v = 1$ to $v = 2$
2. Change in vibrational energy =
   $\Delta E_{1 \to 2}$

3. The vibrations are harmonic, therefore the energy levels are equally spaced
$\therefore \quad \Delta E_{4 \to 3} = \Delta E_{1 \to 2}$
4. Therefore the vibration to vibration transfer is *resonant*, it is also *fast*.
5. And there is no excess energy to be transferred to translation.

All other exchanges can be treated in a similar fashion, showing them to be resonant and fast.

Intermolecular vibration to vibration transfers bring the molecule rapidly down to the first excited level from which it moves, in the rate-determining step, to the ground state by vibration to translation transfer.

Rotational energy changes must also be included, and these contribute in a way which facilitates energy transfer. Rotational contributions can increase the number of possible resonant transfers, or can change a near resonant transfer into a faster resonant transfer as shown in Figure 9.7 and discussed below.

1. *Process 1*: a resonant transfer

Changes in molecule 1 on collision with molecule 2

1. Changes quantum state from
   $v = 3, J = 0$
   to $v = 1, J = 0$
2. Change in energy = $\Delta E_1'$

Changes in molecule 2 on collision with molecule 1

1. Changes quantum state from
   $v = 0, J = 5$
   to $v = 2, J = 4$
2. Change in energy = ΔE_1

3. But $\Delta E_1' = \Delta E_1$
4. and so the vibration to vibration transfer is *resonant*, it is also *fast*.
5. And there is no excess energy to be transferred to translation.

2. *Process 2*: a resonant transfer

The energy change in molecule 1 is equal to the energy change in molecule 2, and so the transfer of vibrational energy is again resonant and fast.

3. *Process 3*: a non-resonant transfer

Changes in molecule 1 on collision with molecule 2	*Changes in molecule 2 on collision with molecule 1*

1. Changes quantum state
from $v = 3, J = 4$
to $v = 2, J = 3$
2. Change in energy is $\Delta E_3'$

1. Changes quantum state
from $v = 1, J = 0$
to $v = 3, J = 1$
2. Change in energy is ΔE_3

    3. But $\Delta E_3' \neq \Delta E_3$
    4. And so the vibrational to vibrational transfer is *non-resonant*, it will
       be *slow*
    5. and if it occurs, there will be a large amount of energy to be
       transferred from translation.

And so, provided the vibration to vibration exchanges are resonant, large amounts of energy can be transferred in this way.

9.6   Intra and intermolecular vibration to vibration transfers in polyatomic molecules

The normal modes of vibration for polyatomic molecules combine together to give various energy states. It is from these states that vibrational energy can be transferred, either in an intramolecular or an intermolecular process.

9.6.1 Intramolecular relaxation

In an intramolecular vibrational energy transfer process the quantum numbers of at least two vibrational modes must alter. For instance the quantum state characterised by (210) could change to a quantum state characterised by (101) which means that mode 1 loses energy, mode 2 loses energy and mode 3 gains energy and a vibration to vibration energy transfer is taking place. If the energies of the two quantum states lie close together, then the transfer will be near resonant and fast. But if the energies of the two quantum states are well separated, then the transfer will be non-resonant and slow. This sort of intramolecular transfer can occur, but it is not the most efficient procedure for causing vibration to vibration transfer. To attain this, intermolecular transfer must be utilised.

9.6.2 Intermolecular relaxation

Opportunities for resonant and near resonant transfers become considerable when intermolecular vibrational energy transfer occurs. This is the main route for relaxation in polyatomic molecules which occurs by simple vibration to vibration transfers, or by simultaneous rotational energy transfer. Rotation greatly increases the number of resonant transitions, and so greatly facilitates vibrational energy transfer.

This mechanism of relaxation is highly efficient as large amounts of energy can be transferred in intermolecular vibration to vibration transfers which are fast, Figure 9.8.

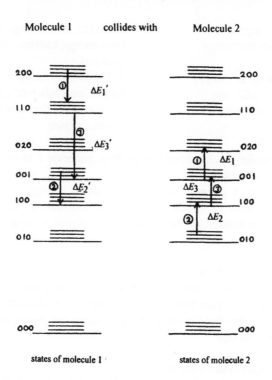

Figure 9.8 Resonant and non-resonant vibration to vibration energy transfers in a polyatomic molecule with rotational contributions.

The requirement for energy transfer is that energy is taken from one energy state in molecule 1 and given to an energy state in molecule 2. These processes are summarised below.

9.6.3 Intermolecular vibration to vibration transfers in polyatomic molecules : Figure 9.8

1. *Process 1*: a resonant transfer

Changes in molecule 1 on collision with molecule 2

Changes in molecule 2 on collision with molecule 1

1. Changes quantum state
from (200), $J = 2$
to   (110), $J = 3$

1. Changes quantum state
from (001), $J = 0$
to   (020), $J = 1$

2. Mode 1 loses energy, mode 2 gains energy, rotational energy is gained.

3. The overall change in energy for this molecule = $\Delta E_1'$, and this is equal to the vertical distance drawn.

2. Mode 2 gains energy, mode 3 loses energy, rotational energy is gained.

3. The overall change in energy for this molecule = ΔE_1, and this is equal to the vertical distance drawn.

4. $\Delta E_1' = \Delta E_1$

and the transfer of vibrational energy from molecule 1 to molecule 2 is **resonant**, and is **rapid**.

5. As this is a resonant transfer, vibrational energy released from molecule 1 goes to vibrational energy in molecule 2; none of it is transferred to translation, but some is transferred to rotation in both molecules.

2. *Process 2*: a resonant transfer

Here molecule 1 changes quantum state from (001) $J = 2$ to (100), $J = 0$ while molecule 2 changes quantum state from (010), $J = 2$ to (100) $J = 1$. A similar argument to that given for process 1 identifies the energy changes occurring, and shows that the overall change is resonant and fast.

3. *Process 3*: a non-resonant transfer

Changes in molecule 1 on collision with molecule 2

Changes in molecule 2 on collision with molecule 1

1. Changes quantum state from (110), $J = 0$ to (001), $J = 0$

2. Mode 1 loses energy, mode 2 loses energy, mode 3 gains energy, no rotational energy change.

3. The overall change in energy for this molecule = $\Delta E_3'$, and this is equal to the vertical distance drawn.

1. Changes quantum state from (100), $J = 0$ to (001), $J = 1$

2. Mode 1 loses energy, mode 3 gains energy, rotational energy is gained.

3. The overall change in energy for this molecule = ΔE_3, and this is equal to the vertical distance drawn.

4. $\Delta E_3' > \Delta E_3$

and the transfer of vibrational energy from molecule 1 to molecule 2 is **non-resonant**, and is **slow**.

5. And so not all of the vibrational energy released from molecule 1 goes to vibration and rotation in molecule 2. Some of it goes to translation.

9.7 Mechanism of relaxation

A pattern of relaxation is now apparent. In the upper levels of the molecule, relaxation occurs by fast resonant and near resonant vibrational to vibrational energy changes involving both *intra* molecular and *inter*-molecular processes. These changes will almost certainly have contributions from rotation. These energy changes are very fast, and give a preferred route over the slower - yet still fast - vibration to translation transfers. When the lowest excited vibrational state is reached the only route possible is deactivation by vibration to translation. This deactivation from $v = 1$ to $v = 0$ must be vibration to translation. If it were vibration to vibration with the energy exchange being $v = 1$ to $v = 0$ in one molecule accompanied by $v = 0$ to $v = 1$ in another molecule then the transfer would be resonant and rapid but it *would not* constitute deactivation and activation in the overall gas. The gap between this first excited state and the ground state is generally bigger than any of the higher quantum gaps so that there is a bottleneck before the final rate-determining step of vibrational to translational energy transfer takes the molecule from the lowest lying excited vibrational state for a diatomic molecule, or from the lowest lying vibrational mode of a polyatomic molecule to the ground state.

The process of activation is the exact reverse. There is a slow rate determining step of activation to the first excited vibrational level, followed by rapid vibration to vibration transfers as the vibrational ladder is ascended.

9.8 Single, double and multiple relaxation times

9.8.1 A single relaxation time
Single relaxation occurs when there is a large gap between the ground level and the first excited state of the lowest mode. This gap *must be large* compared with the gap between the first excited state mode and the next one up in energy, Figure 9.9.

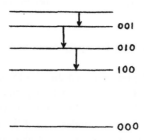

Figure 9.9 A single relaxation time for a polyatomic molecule

In the common single relaxation mechanism immediately above, vibrational to vibrational energy transfer takes the molecule to the lowest lying mode (100), from which it moves to the ground state (000) by the slow vibration to translation transfer. The lowest lying state of mode 2 is (010), and of mode 3, (001). These lie above the lowest lying state of mode 1, (100). The final step is not from (010), or (001), to (000), because the energy which would have to be transferred to translation for these transfers is larger in both cases than is that transferred to translation by relaxation from (100). The molecule moves to the ground state by the fastest of the three possible routes for vibration to translation transfer, that is by (100) → (000), rather than by the slower (010) → (000), or the even slower (001) → (000). Modes 2 and 3 deactivate from their lowest lying states by vibration to vibration transfer; (001) moving to (010) which, in turn, goes to (100), and transfer to translation then occurs.

9.8.2 Two relaxation times

Some molecules, SO_2, CH_2Cl_2, C_2H_6 have the large gap in energy between the two lowest lying modes. This allows two independent mechanisms for relaxation, Figure 9.10

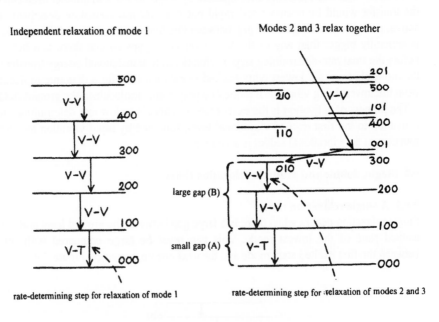

Figure 9.10 A double relaxation time for a polyatomic molecule.

In this case, the large gap (B) lies between the two lowest lying modes, mode 1 lying at (100) and mode 2 lying at (010). The lowest lying state of mode 3 is at (001) which lies above both (010) and (100). The lowest lying mode (here mode 1) relaxes independently, while the two higher, modes 2 and 3, relax together by vibrational-vibrational energy transfer until the molecule reaches state (010). Deactivation then takes the molecule from state (010) to state (100) by vibration to vibration transfer between modes 2 and 1, from which the normal vibration to translation transfer, (100) → (000), can occur. The state (010) lies well above the lowest excited mode (100), and has an energy gap to (100) much larger than that between (100) and (000).

The rate-determining step in energy transfer corresponds to the biggest energy gap, which is that between (010) and (100), made up of two steps (010) → (200) followed by (200) → (100). Both these steps involve vibration to vibration transfers. Since the process (010) → (200) involves a change in quantum number for two different modes, the rate determining step for relaxation of modes 2 and 3 is (010) → (200) followed by the relatively slow, but not rate-determining, vibration to translation transfer (100) → (000).

Mode 1 relaxes independently by fast vibration to vibration transfer for the upper levels, followed by the usual rate-determining vibration to translation transfer, (100) → (000) from the lowest excited state of mode 1.

Because there are two rate-determining steps there are two relaxation times, one for relaxation of mode 1, and the other for modes 2 and 3 relaxing independently of mode 1.

If several modes relax independently there is multiple relaxation, with the number of relaxation times being the same as the number of independent routes. This is rarely observed.

9.9 Energy transfer results relevant to kinetics

9.9.1 Vibration to translation transfer
Although the preferred relaxation in upper levels is by vibration to vibration, energy transfer involving vibration to translation can still be studied for higher levels. Table 12.5 shows data for vibration to translation transfers in HF, and confirms the prediction that the rate constant for transfer increases as the vibrational ladder is ascended.

Table 9.5   Vibrational relaxation of HF

from	Temp/K	$10^{12}k/cm^3s^{-1}$
$v = 1$	298	1.63
$v = 2$	296	16.4 ± 0.5
$v = 3$	296	26.0 ± 0.5
$v = 4$	296	27.0 ± 0.5
$v = 5$	296	8.6 ± 0.5

9.9.2 Comparison of rate constants for vibration-vibration transfer with vibration-translation transfer
Theory predicts that the rate determining step in vibrational activation and deactivation is a vibration - translation transfer between the zero'th and first excited vibrational states. Observation confirms this prediction. For instance the vibration to vibration transfer (011) to (100) for NOCl has a rate constant $k = (4 \pm 1) \times 10^{-11}$ molecule^{-1} cm^3 s^{-1} which is a very rapid rate corresponding to approximately four collisions per transfer. This can be compared with the rate-determining vibration to translation transfer (001) to (000) which has the lower rate constant of $k = 5.6 \times 10^{-12}$ molecule^{-1} cm^3 s^{-1}. Likewise the vibration to vibration transfer for CH_4, (0010) to (0001) has a rate constant of 70×10^{-13} molecule^{-1} cm^3 s^{-1} which can be compared with the vibration to translation transfer with rate constant equal to 0.22×10^{-13} molecule^{-1} cm^3 s^{-1}. It is immediately clear that, yet again, vibration to vibration is the most efficient process.

9.9.3 Vibration to vibration transfer in CO_2
Typical vibration to vibration transfer occurs with a wide range of rates, and there is evidence that resonant and near resonant transfers are more efficient, and have higher rate constants than non-resonant transfers.

The lower lying energy states for CO_2 are shown in Figure 9.11, and a selection of rate constants for vibration to vibration transfers are given in Table 9.6. These can be compared with the very much slower vibration to translation transfer taking the lowest lying mode to the ground state, with rate constant equal to 6.5×10^{-15} molecule^{-1} cm^3 s^{-1}.

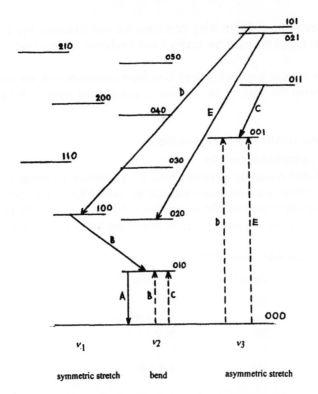

symmetric stretch   bend   asymmetric stretch

Full arrows indicate changes undergone by first molecule

Broken arrows indicate changes undergone by second molecule

Figure 9.11 Energy states for CO_2, showing transitions given in Table 9.6.

Table 9.6 Vibrational to vibrational transfers in CO_2 showing a wide range in rates

transition	process	$\dfrac{k}{cm^3 s^{-1}}$	Z
A	$CO_2(010)+CO_2 \rightarrow CO_2(000)+CO_2$ (vibration to translation)	6.5×10^{-15}	$\sim 2 \times 10^4$
B	$CO_2(100)+CO_2(000) \rightarrow CO_2(010)+CO_2(010)$	3.1×10^{-13}	~ 500
C	$CO_2(011)+CO_2(000) \rightarrow CO_2(001)+CO_2(010)$	1.6×10^{-10}	~ 1
D	$CO_2(101)+CO_2(000) \rightarrow CO_2(100)+CO_2(001)$	1.3×10^{-10}	~ 1
E	$CO_2(021)+CO_2(000) \rightarrow CO_2(020)+CO_2(001)$	1.2×10^{-10}	~ 1

These transitions are marked in Figure 9.11

9.9.4   Anharmonicity effects in the vibrational to vibrational energy transfer in CO

This study is interesting because it requires a much finer and more detailed description of the mechanism of vibrational to vibrational energy transfer.

The relaxation rate normally increases as the vibrational quantum number increases, so that higher levels will show greater rates of deactivation than lower levels. The data in Table 9.7 do not show this trend - here the relaxation rates first increase, and then decrease with increasing quantum number. This shows conclusively that anharmonicity of the vibration must be considered in any prediction of the effect of vibrational quantum number, v, on rate.

Table 9.7   Vibrational-vibration energy transfer

$CO\ (v = n) + CO\ (v = 0) \quad \rightarrow \quad CO\ (v = n\text{-}1) + CO\ (v = 1)$

			$10^3 k/cm^3 s^{-1}$ at	
n	75K	100K	300K	500K
2	71	40	19.3	14.5
3	124	34	22.7	18.0
4	37	10	21.8	20.5
5	14	3.1	15.6	18.7
6	6.2	0.93	8.7	14.2
7	2.8	0.25	3.9	8.8
8	1.2		2.0	5.1
9			1.2	2.8
10			0.75	1.7
11			0.47	1.1
12				0.5

In a harmonic oscillator all vibrational-vibrational energy exchanges are resonant, whereas in the anharmonic oscillator transfer is not resonant, except in the very low levels where it is approaching near resonance. Resonant and near resonant transfers are faster than non resonant and, if anharmonicity is important, this should result in the rate constants showing the normal increase in low levels, but moving over to a decrease in rate at higher levels. This is precisely what is observed. The relaxation rate constants, therefore, have two contributing factors:

(a) an increase in rate as v increases,

(b) a decrease in rate as anharmonicity increases and non-resonance becomes effective, with a balance at a certain value of v.

Table 9.7 shows that except for 100 K a maximum in rate constant is found as v increases.

If the only factor determining the relaxation is the dependence of rate constant on the vibrational quantum number, then the rate constant would be proportional to v. By dividing the rate constant by the value of v, the resulting values should reflect the effect of anharmonicity. This is given in Table 9.8. The maxima in rate constant at low v are now removed, and the decreasing values of the rate constant as the vibrational ladder is ascended reflects the increasing anharmonicity. This results in a decrease in the number of possible resonant transfers, and hence a decrease in the rate constant for energy transfer.

Table 9.8 Analysis of vibrational-vibrational energy transfer for CO

n	75K $\dfrac{10^{13} k/cm^3 s^{-1}}{n}$	100K $\dfrac{10^{13} k/cm^3 s^{-1}}{n}$	300K $\dfrac{10^{13} k/cm^3 s^{-1}}{n}$	500K $\dfrac{10^{13} k/cm^3 s^{-1}}{n}$
2	35.5	20	9.65	7.25
3	41	11.3	7.57	6.00
4	9.25	2.5	5.45	5.12
5	2.8	0.62	3.12	3.74
6	1.03	0.16	1.45	2.37
7	0.40	0.04	0.56	1.26
8	0.15		0.25	0.64
9			0.13	0.31
10			0.075	0.17
11			0.043	0.10
12				0.06

9.9.5 Vibrational energy exchange with two relaxation times

All of the systems so far discussed have one relaxation time. SO_2 on the other hand, shows two relaxation times indicating a double mechanism for relaxation, with two of the modes relaxing together and the third mode relaxing independently, Figure 9.12. Transfer of energy between the two stretching modes is easy, but transfer between bending and stretching is difficult.

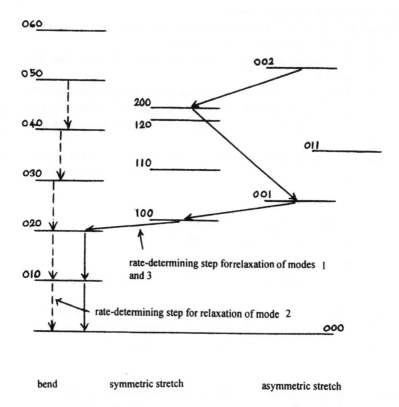

bend   symmetric stretch   asymmetric stretch

Sequence of transfers which does not involve any
change in energy in ν_1, ν_3. This gives relaxation
pattern for ν_2 with rate-determining step as
indicated

Sequence of transfers which does not involve any
change in energy in ν_2 This gives relaxation
pattern for ν_1, ν_3 with rate-determining step as
indicated

Figure 9.12   Relaxation of SO$_2$, two relaxation times.

The two stretching modes relax together by exchange of vibrational energy between
them until the lowest lying mode of the pair is reached. This is the symmetric stretch
(100). There is now a rate-determining step of vibration to vibration transfer from this
mode into the second harmonic of the bending mode v_2. The (020) state then relaxes by
a rapid resonant vibration to vibration transfer to the lowest excited state (010), from
which a slow process of vibration to translation takes the molecule to the ground state.
This last step, although slow, is still faster than the rate determining step of vibrational
transfer from the symmetric stretch to the bending mode (100) → (020).

The bending mode relaxes independently with the usual rate determining step of
transfer of vibration to translation (010) → (000).

The rate determining step for relaxation by vibration to translation has a rate constant
approximately equal to 7.8×10^{-13} molecule^{-1} cm^3 s^{-1}, while the rate determining step for
vibration to vibration transfer between the asymmetric stretch and the bending mode has
a rate constant approximately equal to 1.3×10^{-13} molecule^{-1} cm^3 s^{-1}. This confirms that
the vibration to vibration rate-determining step is slower than the rate-determining
vibration to translation step found for the independent relaxation of the bending mode.

9.10 Transfer of rotational energy
This transfer was formerly studied by ultrasonic methods but these have now been superseded by laser and molecular beam studies.

Transfer of rotational energy requires far fewer collisions than does vibrational transfer, Table 9.4.

Rotation to translation transfer is very rapid, and multiple quantum jumps can occur. However, the rate of transfer decreases as the value of J increases, and this in direct contrast to the increasing rate of transfer which occurs on climbing the vibrational ladder.

Rotation to rotation transfers are nearly always non-resonant and slow, and this can be contrasted with the corresponding situation in vibration where resonant transfers are fast.

9.10.1 Rotational energy changes in HF
A laser study has been carried out for the process

$$HF (v = 1, J = n_1) + HF (v = 0, J \text{ unspecified})$$
$$\rightarrow HF (v = 1, J = n_2) + HF (v = 0, J \text{ unspecified})$$

Table 9.9 show the results of this laser study, and these confirm the prediction that rate constants for a given initial J decrease as the change in J increases, and larger amounts of rotational energy are transferred at a lower rate, Figure 9.13.

Table 9.9 Rate constants for rotational energy transfer for HF

initial $J = n_1$	final $J = n_2$	$10^6 k/\text{mmHg}^{-1}\text{s}^{-1}$
2	3	22
2	4	6.3
2	5	1.6
2	6	0.42
2	7	0.083
3	4	6.0
3	5	1.4
3	6	0.42
3	7	0.12
3	8	0.02
4	5	3.4
4	6	0.76
4	7	0.21
4	8	0.017
5	6	1.5
5	7	0.23

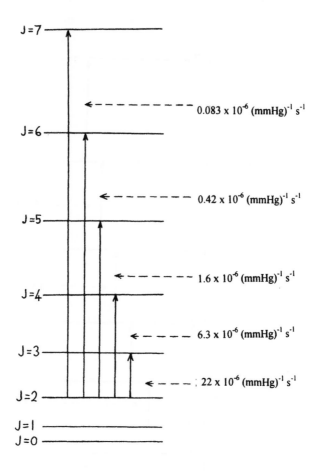

Figure 9.13 Rotational energy transfer in HF - strong collisions, that is multiple quantum jumps.

Likewise, jumps of a given ΔJ correspond to different energies and depend on the initial J. The larger the gap, the bigger the energy transfer and the slower is the rate, Figure 9.14.

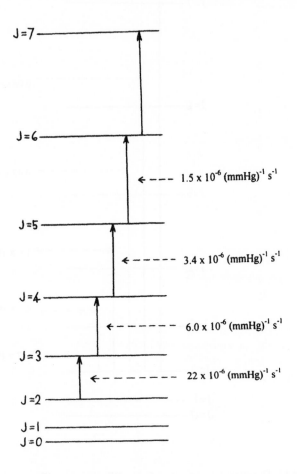

Figure 9.14 Rotational energy transfer in HF showing the decrease in rate constant as *J* increases for ladder climbing, that is single quantum jumps.

10

Molecular Beam Studies of Reactive Scattering

In Chapters 7 and 8 scattering on collision was discussed in terms of cross sections, rate constants and transition probabilities. In this chapter these results will be discussed in terms of contour diagrams.

10.1 Contour diagrams

The amount of scattering and the distribution of velocities of the products are studied as a function of scattering angle. By appropriate conversions these can be neatly displayed on a contour diagram given in terms of the centre of mass of the reacting system. The contours join up regions of equal scattering, and so the diagram shows immediately the distribution of scattering in the products.

In a centre of mass representation such as the contour diagram, A and BC are approaching each other along their line of centres, and the origin of the diagram is the centre of mass of the system, Figure 10.1.

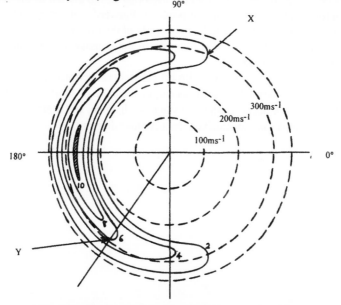

Figure 10.1 A typical diagrammatic illustration of results of a molecular beam experiment showing scattering and velocity distribution.

Any point on this diagram represents a velocity at a given angle of scattering, and for a particular amount of scattering. For instance, at point X, the velocity of the products scattered into that point is 300 m s^{-1}, the angle of scattering is 69° and there is only a small amount of scattering, given by the value of the contour, 2. Contours join up regions of equal scattering, while concentric circles represent regions where the scattered particles have equal velocities, for example all particles scattered into points lying on the shaded area have a velocity of around 270 m s^{-1}. Any points lying on the line have been scattered at an angle of 234°.

The contour diagram can be "read" as follows.

Point X represents

an amount of scattering	$\propto$ 2
scattering into angle	= 69°
with a velocity	= 300 m s^{-1}

Point Y represents

an amount of scattering	$\propto$ 6
scattering into angle	= 234°
with a velocity	= 293 m s^{-1}

In this diagram there is very little scattering into the two right hand quadrants shown by the low lying contours. The highest amount of scattering lies in the two left hand quadrants where the contours are high. The most probable scattering lies within the shaded contour, 10, between angles 164° and 196°, and with a most probable velocity between 267 and 280 m s^{-1}.

The initial direction of motion with respect to the centre of mass is from left to right. In this diagram the region of most probable scattering lies to the left of the origin, and there is therefore backward scattering of products. Forward scattering would show the region of most probable scattering to lie to the right of the origin.

The most probable velocity is also found from the contour diagram, and is in the region of maximum scattering.

When two molecules collide the collision can last sufficiently long, $> 10^{-13}$s, for several rotations and many vibrations to occur, or it can be of short duration, $< 10^{-13}$s, so that rotation does not occur. Both these situations happen, and molecular beams can demonstrate these situations conclusively on a contour diagram.

If the collision is long and several rotations occur, the molecules will forget in what directions they ought to part company, and instead shoot off in random directions. This results in a contour diagram which is symmetrical about the centre of mass, Figure 10.2.

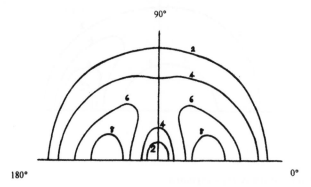

Figure 10.2 A symmetrical contour diagram for a reaction with a collision complex.

On the other hand, if the collision is of very short duration there will be no time for rotation, and the molecules will not have time to forget the direction in which they ought to move. The contour diagram is now asymmetric about the centre of mass, Figures 10.1, 10.3, 10.4 and 10.5.

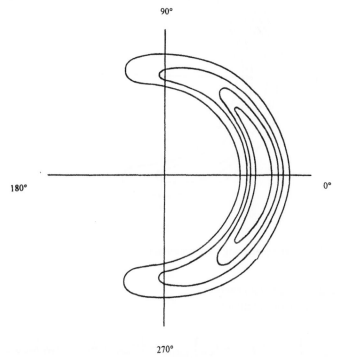

Figure 10.3 Forward scattering - asymmetrical.

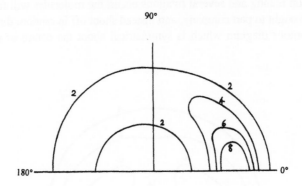

Figure 10.4 Forward scattering - asymmetrical.

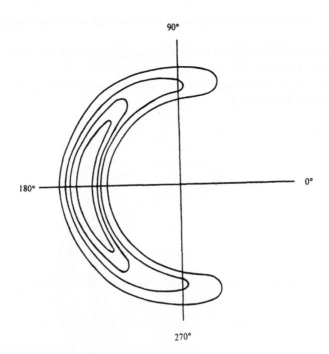

Figure 10.5 Backward scattering - asymmetrical.

10.2 Energy changes on reaction

The energy change on reaction is

$$-\Delta E_{\text{reaction}} = \left\{\varepsilon_{\text{trans}} + \varepsilon_{\text{rot}} + \varepsilon_{\text{vib}}\right\}_{\text{products}} - \left\{\varepsilon_{\text{trans}} + \varepsilon_{\text{rot}} + \varepsilon_{\text{vib}}\right\}_{\text{reactants}} \tag{10.1}$$

$\Delta E_{\text{reaction}}$ can be found by experiment. Translational, rotational and vibrational energy can be selected for the reactants, and the translational energy of the products found from the time of flight mass spectrometer. This would give enough information to solve Equation 10.1 to give the internal energy of the products.

More detailed experiments would select given specific quantum states for the reactants, and if the populations of the selected vibrational and rotational states are determined then the initial distribution of energy between the rotational and vibrational states and translation will be known.

Determination of all vibrational states and their populations, and of all rotational states and their population taken together with the distribution of translational energy will give the distribution of energy in the products. This gives sufficient information to enable determination of the following.

(i) The way in which energy is channelled into vibrational energy can be found from the ratio

$$\frac{\varepsilon_{\text{vib}}}{\varepsilon_{\text{vib}} + \varepsilon_{\text{rot}} + \varepsilon_{\text{trans}}}$$

which gives the fraction of the total energy of the products which appears as vibrational energy.

(ii) The corresponding ratio

$$\frac{\varepsilon_{\text{rot}}}{\varepsilon_{\text{vib}} + \varepsilon_{\text{rot}} + \varepsilon_{\text{trans}}}$$

gives the fraction of the total energy of the products which appears as rotational energy, and so gives information as to how energy is channelled into rotational energy in the products.

(iii) Comparison of the initial quantum states and populations of the reactants, with the final quantum states and their populations shows how the initial distribution of energies is redistributed as a result of reaction.

10.3 Forward scattering - the stripping or grazing mechanism
In a molecular beam experiment, results for the reaction

$$A + BC \quad \rightarrow \quad AB + C$$

can be summarised in a contour diagram such as Figures 10.3 and 10.4.

The product AB is shown to have forward scattering with the maximum very close to $0°$, so that A is not deflected much during the reactive collision. Reactions with forward scattering have large impact parameters, b, and large cross sections, σ. AB emerges from the collision, and continues only slightly deflected from the direction of A before collision. Likewise, C is not much affected by the collision, and continues in its original direction only slightly deflected.

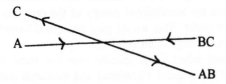

The mechanism is called "stripping" because A simply strips B off BC in passing and from a distance, or put alternatively, A can be seen as taking a distant sideways swipe at BC so that B is pulled off from BC, leaving C to continue in almost the same direction. A, having made a successful grab at B, continues on as AB, with little change of direction. The lack of deflection shows that the interaction between A and BC is strong so that there can be significant interactions at a distance, thereby removing the need for close contact. The cross section for reaction is large, which is consistent with this interpretation. The most probable translational energy of the products is low, so that most of the energy of reaction goes into vibrational excitation in the products.

Trajectory calculations on assumed potential energy surfaces predict this sort of behaviour when the surface has an early energy barrier in the entrance valley - an attractive potential energy surface. This would imply a large cross sectional area for reaction. The calculations also predict that a large fraction of the reaction energy will be found as vibration in the products and that translational energy will be more important in promoting reaction. Excess translational energy in the reactants should appear as excess translational energy in the products. Molecular beam experiments with forward scattering confirm these predictions.

One of the first reactions shown to occur with forward scattering is the reaction between atomic potassium and molecular iodine.

$$K^{\cdot}_{(g)} + I_{2(g)} \quad \rightarrow \quad KI_{(g)} + I^{\cdot}_{(g)}$$

Other reactions of this type include alkali metals with halogen molecules, and atomic oxygen reactions such as

$$O^{\cdot}_{(g)} + CS_{2(g)} \quad \rightarrow \quad OS_{(g)} + CS_{(g)}$$

The "stripping" mechanism is also favoured by many ion-molecule reactions.

10.4 Backward scattering - the rebound mechanism
The molecular beam experimental results for a reaction

$$P + MN \quad \rightarrow \quad PM + N$$

are summarised in a contour diagram, such as Figure 10.5.

The product PM is shown to have maximum scattering very close to 180° so that P must rebound almost totally backwards from the reactive collision. This is interpreted in terms of a head-on collision which occurs at a very close encounter, giving a small impact parameter and small cross section. The impact parameters associated with

backward scattering are all much smaller than those for stripping reactions, being more
in line with hard sphere values.

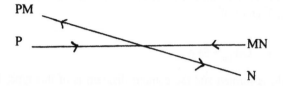

This can be explained by suggesting that P hits MN head-on, pulls M away from N
and, under the influence of the repulsive forces between M and N, PM moves almost
totally backwards. N also rebounds backwards from PM under the same repulsive
forces, and returns back only slightly deflected from its original path.

In most rebound reactions the energy of reaction is channelled into translation in the
products, showing that they are separating under short range repulsions. This is totally
different from the "stripping" mechanism where the reactive collision takes place
under a strong attractive interaction, and the energy of reaction is channelled into
vibration.

Again trajectory calculations predict this sort of behaviour for a potential energy
surface with a late barrier in the exit valley - a repulsive potential energy surface. This
would imply small impact parameters and cross sectional areas for reaction. The
calculations also predict that a large fraction of the reaction energy is channelled into
translational energy of products, but that vibrational energy is more effective in
promoting reaction. They also predict that excess vibrational energy in reactants
appears as excess vibrational energy in the products. Again molecular beam
experiments with backward scattering confirm these predictions.

The most quoted reaction proceeding by a rebound mechanism is

$$K^{\cdot}{}_{(g)} + CH_3I_{(g)} \quad \rightarrow \quad KI_{(g)} + CH_3{}^{\cdot}{}_{(g)}$$

There are many other typical reactions such as those of other alkali metals with CH_3I,
reactions of atomic hydrogen and deuterium with halogens, and methyl halogen
reactions.

10.5 Reverse reactions for reactions associated with forward and backward
  scattering

For both types of reaction, features of the reverse reactions can be inferred.

If the forward step of a reversible reaction is associated with forward scattering, then
the reverse step will be associated with backward scattering, small impact parameters
and cross sections, promotion of reaction by vibrational energy and channelling of
reaction energy into translational energy in the products.

Likewise, if the forward step of a reversible reaction is associated with backward
scattering, then the reverse step will be associated with forward scattering, large impact
parameters and cross sections, promotion of reaction by translational energy and
channelling of reaction energy into vibration in the products.

10.6 Long-lived complexes

When the lifetime of the collision is of the order of several rotations, a collision complex is formed, and the contour diagram can be symmetric with respect to 90°, with maxima appearing at 0° and 180°.

In the reaction

$$Cs^{\cdot}_{(g)} + SF_{6(g)} \quad \rightarrow \quad CsF_{(g)} + SF_{5(g)}^{\cdot}$$

a collision complex $CsSF_6$ is formed and the contour diagram is of this type, Figure 10.6.

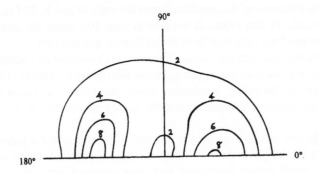

Figure 10.6 Contour diagram showing the collision complex for the reaction
$$Cs^{\cdot}_{(g)} + SF_{6(g)} \quad \rightarrow \quad CsF_{(g)} + SF_{5}^{\cdot}_{(g)}$$

In this type of reaction, interactions keep the two reactants in close contact with each other for long enough for them to forget the directions in which they would have parted had they been able to do so before rotation. They, therefore, move apart in random directions.

In the reaction

$$F^{\cdot}_{(g)} + C_2H_{4(g)} \quad \rightarrow \quad C_2H_3F_{(g)} + H^{\cdot}_{(g)}$$

the contour diagram has the maximum at 90°, Figure 10.7.

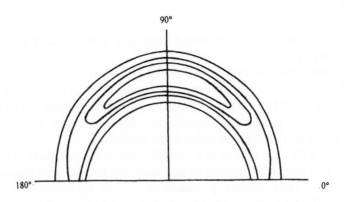

Figure 10.7 Contour diagram showing the collision complex for the reaction
$$F^{\cdot}_{(g)} + C_2H_{4(g)} \quad \rightarrow \quad C_2H_3F_{(g)} + H^{\cdot}_{(g)}$$

while the reaction

$$Cl\cdot_{(g)} + CH_2\!=\!CHBr_{(g)} \quad \rightarrow \quad CH_2\!=\!CHCl_{(g)} + Br\cdot_{(g)}$$

gives a diagram showing symmetrical maxima around 90°, Figure 10.8.

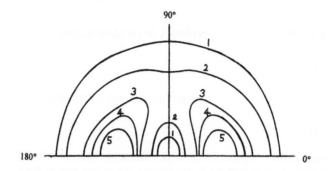

Figure 10.8 Contour diagram showing the collision complex for the reaction
$$Cl\cdot_{(g)} + CH_2 = CHBr_{(g)} \quad \rightarrow \quad CH_2 = CHCl_{(g)} + Br\cdot_{(g)}$$

This almost perfect symmetry can be compared with the lack of symmetry shown for the reaction

$$Cl\cdot_{(g)} + CH_2\!=\!CHCH_2Br_{(g)} \quad \rightarrow \quad ClCH_2CH\!=\!CH_{2(g)} + Br\cdot_{(g)}$$

where the diagram is displayed around 90° with maxima on each side, but with differing intensities, Figure 10.9.

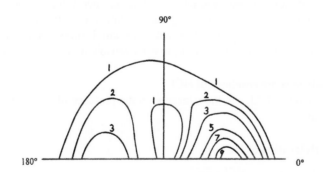

Figure 10.9 Contour diagram showing a collision complex for the reaction
$$Cl\cdot_{(g)} + CH_2 = CH\,CH_2Br_{(g)} \quad \rightarrow \quad CH_2 = CH\,CH_2Cl_{(g)} + Br\cdot_{(g)}$$

This particular "symmetric" contour diagram is believed to be associated with a shorter lived complex which survives only one or two rotations.

Reactions involving collision complexes are associated with wells on the potential energy surface.

10.7 Steric effects

Polar molecules can be aligned using crossed electric fields. If the polarity of the field is altered the alignment of the molecules will change. When this is done for the reactions of K and Rb with CH_3I the reaction cross sections are different for the two polarities. The larger cross section is attributed to direct attack of K on to the I in CH_3I, while the smaller cross section is associated with the K approaching the CH_3 hindered side.

$CH_3I \leftarrow K$ preferred, larger cross section

$$K \rightarrow H - \overset{\displaystyle H}{\underset{\displaystyle H}{C}} - I$$ less preferred, smaller cross section

When this experiment is repeated with CF_3I, different behaviour is found.

When the polarity is such that the K approaches the iodine end of the CF_3I, product KI is scattered backwards with the low impact parameters typical of a rebound reaction, believed to be the result of K hitting CF_3I head-on on the iodine.

When the polarity is reversed and the K approached the CF_3 end of the CF_3I, product KI is now scattered forwards with the large impact parameter typical of a stripping reaction, believed to be a result of a potassium atom flying past the CF_3I and extracting the iodine atom.

The conclusion is that the iodine atom is less shielded in CF_3I than in CH_3I.

More work of this type should help to define the geometry of reactions, and this should help in the determination of potential energy surfaces.

10.8 State to state kinetics

Determination of cross sections and rate constants for energy transfer and reaction is one ultimate aim in this fundamental approach to kinetics. In this respect molecular beam studies are still a young discipline, but, with time it should be possible to have fairly extensive tabulated data comparable to that available from macroscopic kinetics.

10.8.1 A study of some reactions of HCl

The self relaxation of HCl, and the relaxation in the presence of other gases has been studied in detail.

Table 10.1 Relaxation of HCl

$HCl\ (v) + HCl \quad \rightarrow \quad HCl + HCl$

$v = 2 \qquad\qquad k = 3.3 \times 10^{-12}\ cm^3 s^{-1} molecule^{-1}$
$v = 1 \qquad\qquad k = 2.6 \times 10^{-14}\ cm^3 s^{-1} molecule^{-1}$

$HCl\ (v) + Br \cdot \quad \rightarrow \quad HCl + Br \cdot$

$v = 2 \qquad\qquad k = 1.8 \times 10^{-12}\ cm^3 s^{-1} molecule^{-1}$
$v = 1 \qquad\qquad k = 2.8 \times 10^{-13}\ cm^3 s^{-1} molecule^{-1}$

Table 10.1 gives a few typical data which show up some of the general trends. When HCl is in the states $v = 1$ and $v = 2$ rotationally enhanced transfer of vibration to translation takes place, and this is borne out by many other studies. In the presence of Br atoms relaxation is still by rotationally enhanced vibration, but when HCl is in $v = 2$, relaxation becomes mixed up with reaction with Br.

$$HCl_{(g)} + Br\cdot_{(g)} \quad \rightarrow \quad HBr_{(g)} + Cl\cdot_{(g)}$$

This reaction has also been studied in an attempt to disentangle the relaxation data.

Various excited states of HCl can be selected and the overall rate constants found for each state, Table 10.2.

Table 10.2 Reaction of Br with HCl

v for HCl	0	1	2	3	4
$k/dm^3mol^{-1}s^{-1}$	1.0×10^{-2}	2.0×10^4	1.0×10^9	7.0×10^9	1.0×10^{10}

These are not true state to state rate constants because they are not between the specified initial states to any one specified final state. However, they do enable an estimate of the effect of vibrational energy on reaction to be made. When $v = 1$ and $v = 2$ there is a considerable enhancement of the rate over $v = 0$, but higher vibrational energies do not lead to much increase over the $v = 2$ value of k.

10.8.2 Reaction of HCl with K·

Reaction of HCl with K· has been studied by molecular beams and lasers

$$K\cdot_{(g)} + HCl_{(g)} \quad \rightarrow \quad KCl_{(g)} + H\cdot_{(g)}$$

Laser excitation placed HCl into $v = 1$ and the rate of reaction of HCl ($v = 0$) and HCl ($v = 1$) with K· was studied using molecular beams.

(i)   *For HCl (v = 0)* there was no reaction at low translational energies, but when the threshold energy was exceeded reaction occurred.

(ii) *For HCl (v = 1)*, the threshold energy is now exceeded, even at very low translational energies at which there is considerable reaction. However, increasing the translational energy caused little increase in rate.

From this, it was concluded that vibrational energy was more important than translational energy, though the effect is not so striking as in the K· + HF reaction, see below. A potential energy surface has been constructed and shown to have a slightly late barrier. Late barriers are associated with vibrational enhancement, and the relatively small enhancement found in the molecular beam experiments is in keeping with the barrier being only a slightly late one.

10.8.3 Reaction of HF

The great potential of molecular beams is demonstrated by the reactions of HF

$$X\cdot_{(g)} + HF_{(g)} \rightarrow XH_{(g)} + F\cdot_{(g)}$$
$$X\cdot_{(g)} + HF_{(g)} \rightarrow XF_{(g)} + H\cdot_{(g)}$$

where X = Li, Na, K, Ba, Sr, H, Cl, Br.

10.8.4 Reaction with HF and Ba·
The reaction

$$Ba\cdot_{(g)} + HF_{(g)} \rightarrow BaF_{(g)} + H\cdot_{(g)}$$

demonstrates this and the reaction has been the subject of intense experimental interest from which a vast amount of information has emerged.

(i) *Activation, deactivation and reaction in the system Ba· + HF*

Activation - deactivation rate constants for

$$HF_{(g)} + HF_{(g)} \rightarrow HF^*_{(g)} + HF_{(g)}$$

have been found from energy transfer studies on HF/HF collisions where the dependence of rotationally enhanced vibration to translation transfer on v and J is found. More extensive work would include vibration to vibration rate constants so that all transfers in the processes of activation and deactivation are known.

Energy transfer in Ba· and HF mixtures give rate constants for $v = 0$ and $v = 1$, but experiments with higher values of v were complicated by reaction.

In the reaction itself

$$Ba\cdot_{(g)} + HF_{(g)} \rightarrow BaF_{(g)} + H\cdot_{(g)}$$

one experiment showed that reaction with HF in the $v = 0$ state gave BaF with a maximum population in $v = 0$, a slightly smaller population in $v = 1$ and progressively decreasing populations in $v = 2$ to $v = 6$. Another experiment from HF in $v = 1$ gives a totally different vibrational distribution. Levels in BaF from $v = 1$ to 12 are now populated with a maximum at $v = 6$. A third experiment looked at the effect of translational energy on the reaction. With HF as $v = 0$, and with high translational energies the maximum population is still in $v = 0$, but levels up to $v = 10$ are now populated.

These experiments suggest that there is a slight translational enhancement, and some vibrational enhancement, indicative of a late barrier.[1]

Cross sections for the reaction at a translational energy of 6.7 kJ mol^{-1} have been found for reaction of HF in state $v = 0$ to form BaF in a large number of vibrational states, and for reaction of HF in state $v = 1$ to form BaF in an even greater number of vibrational states. Table 10.3 shows the data. The cross sections could be converted into state to state rate constants by integrating over the velocity distribution.

[1] Or a "cutting the corner trajectory", see Chapter 6.

Table 10.3　Cross sections at translational energy 6.7 kJ mol^{-1} for

Ba + HF　→　BaF + H

v in product	0	1	2	3	4	5	6
$10^3\sigma/nm^2$ for $v=0$ in reactant	3.2	3.3	2.1	0.7	0.3	0.1	
$10^3\sigma/nm^2$ for $v=1$ in reactant	3.2	3.3	2.1	2.1	1.85	1.57	1.58

v in product	7	8	9	10	11	12	13
$10^3\sigma/nm^2$ for $v=0$ in reactant							
$10^3\sigma/nm^2$ for $v=1$ in reactant	1.59	1.63	1.18	0.52	0.28	0.17	0.09

The cross sections, and consequently the state to state rate constants decrease as the vibrational energy of the products increases, but after BaF with $v = 3$ is reached the cross sections are consistently higher for HF in $v = 1$ state. However, product formation at $v = 0 - 3$ still dominates the reaction kinetics. This data shows evidence of vibrational enhancement.

An even more exhaustive study, Table 10.4, gave values of the cross section over a wide range of translational energies for reaction of Ba with HF in the $v = 0$ state producing BaF in a large number of vibrational states. These cross sections are equivalent to state to state rate constants over a range of translational energies. The results are in keeping with those in Table 10.3 in that the cross sections at any given translational energy decrease as the vibrational energy of the product increases. The total cross section for the reaction at each translational energy increases with translational energy up to a maximum value. This is indicative of some translational enhancement.

Table 10.4 Cross sections at various translational energies for

$$Ba + HF \longrightarrow BaF + H$$

10^3 x cross section/nm^2 for $v = 0$ in reactant

energy/ kJ mol^{-1}	6.7	13.0	22.2	27.2	37.7	43.9	46.0	55
v in product								
0	3.2	2.14	1.94	1.75	2.17	2.14	2.39	2.34
1	3.3	2.72	2.13	1.85	1.78	1.94	2.11	2.11
2	2.1	3.02	2.15	1.95	1.64	1.58	1.56	1.42
3	0.7	1.63	1.79	1.80	1.26	1.35	1.32	1.35
4	0.3	0.42	1.17	1.36	1.01	1.04	0.88	0.72
5	0.1	0.07	0.55	0.82	0.89	0.75	0.69	0.69
6			0.17	0.33	0.52	0.51	0.39	0.41
7			0.09	0.14	0.34	0.31	0.27	0.36
8					0.16	0.17	0.19	0.19
9					0.09	0.08	0.06	0.12
10					0.08	0.07	0.06	0.08
11					0.05	0.05	0.06	0.06
12								0.05
13								0.05

However, when all the data is taken into consideration, the conclusion is that the increase in the total rate is only slightly greater with increase in vibrational energy than with increase in translational energy. Also the fraction of product energy appearing in vibration is unusually small, most of it appearing as translation. The reaction

$$Ba^{\cdot}_{(g)} + HF_{(g)} \quad \rightarrow \quad BaF_{(g)} + H^{\cdot}_{(g)}$$

is only slightly exothermic, $\Delta H = -18.41$ kJ mol^{-1} and most exothermic reactions show an early barrier in the potential energy surface. This reaction is unusual in that it displays characteristics of an early barrier - translational enhancement, and characteristics of a later barrier - vibrational enhancement, with a high proportion of translational energy in the products. It is thought that this indicates a later barrier than the early one predicted for exothermic reactions. A potential energy surface and trajectory calculations could help to clarify these details.

10.8.5 Other reactions of HF
A number of other reactions of HF also show vibrational enhancement.

$$K^{\cdot}_{(g)} + HF_{(g)} \quad \rightarrow \quad KF_{(g)} + H^{\cdot}_{(g)}$$

gives considerable vibrational enhancement. When HF in the $v = 0$ state is compared with reaction of HF in the $v = 1$ state at low translational energies, the rate increase was vastly greater than that found when there was a comparable change in translational energy. There is therefore considerable vibrational enhancement, and a late barrier is postulated, and is found in a trajectory - potential energy surface calculation.

Other reactions of HF which show vibrational enhancement are

$$Sr\cdot_{(g)} + HF_{(g)} \rightarrow SrF_{(g)} + H\cdot_{(g)}$$
$$Cl\cdot_{(g)} + HF_{(g)} \rightarrow HCl_{(g)} + F\cdot_{(g)}$$
$$Br\cdot_{(g)} + HF_{(g)} \rightarrow HBr_{(g)} + F\cdot_{(g)}$$

Indeed the reaction with Sr occurs only if $v = 1$, and not at all if $v = 0$.

Reactions between light molecules have been intensively studied over the past decade or so. This has probably resulted from the very considerable increases in accuracy of quantum mechanical calculations of potential energy surfaces for these reactions. These have, in turn, enabled accurate trajectory calculations for motion over these surfaces to be made, and the increasing detail of molecular beam experiments allows detailed testing of the predictions of the surface-trajectory calculations to be possible.

10.9 Ion-molecule studies

Ion-molecule reactions in the gas phase have been studied by mass spectrometry, chemi, photo and field ionisation techniques. Table 10.5 lists some typical reactions.

Table 10.5 Ion-molecule reactions

H_2^+	+	H_2	$\rightarrow$	H_3^+	+	H·
H_2^+	+	O_2	$\rightarrow$	HO_2^+	+	H·
N^+	+	O_2	$\rightarrow$	NO^+	+	O·
O^+	+	N_2	$\rightarrow$	NO^+	+	N·
O_2^+	+	NO	$\rightarrow$	NO^+	+	O_2
Kr^+	+	H_2	$\rightarrow$	KrH^+	+	H·
Ar^+	+	H_2	$\rightarrow$	ArH^+	+	H·
$C_2H_4^+$	+	C_2H_4	$\rightarrow$	$C_3H_5\cdot^+$	+	$CH_3\cdot$
CH_4^+	+	CH_4	$\rightarrow$	$CH_5\cdot^+$	+	$CH_3\cdot$
$tC_4H_9\cdot^+$	+	iC_5H_{12}	$\rightarrow$	iC_4H_{10}	+	$tC_5H_{11}\cdot^+$

Ion-molecule reactions are of considerable interest theoretically. When one of the reactants is an ion, attractive interactions can now extend over distances much larger than those implied by a simple collision cross-section. This alters the nature of collision in the ion-molecule collision compared to the molecule-molecule collision.

Several "collision theory" type theories have been proposed. For instance, one allowing for ion-dipole interactions has been derived and compared with a hard sphere collision theory. Not surprisingly, the one allowing intermolecular interactions is more satisfactory.

Many ion-molecule reactions have large cross sections implying that reaction can occur at the large distances apart which are typical of reactions dominated by long range interactions of the ion-dipole, ion-induced dipole type. Ion-molecule reactions do not have a threshold energy, and the cross section decreases with increase in energy, in contrast to typical reactions between neutral molecules which do have a threshold energy and have cross sections increasing to a maximum.

There are two typical ion-molecule mechanisms inferred from molecular beam work.

Direct reaction involves one collision followed by reaction. Direct reaction shows asymmetric scattering of products, often forward peaked and showing the large cross sections corresponding to a stripping action which can occur when the reactants are at large distances apart.

Reactions showing this behaviour include

$$
\begin{aligned}
N_2^+ + H_2 &\rightarrow N_2H^+ + H\cdot \\
CO^+ + H_2 &\rightarrow COH^+ + H\cdot \\
N_2^+ + CH_4 &\rightarrow N_2H^+ + CH_3\cdot
\end{aligned}
$$

and simple reactions of this type are common.

The mechanism going via a collision complex is more common for more complex reactions such as

$$
\begin{aligned}
CH_4^+ + CH_4 &\rightarrow CH_4 + CH_4^+ \\
C_2H_4^+ + C_2H_4 &\rightarrow C_3H_5^+ + CH_3\cdot \\
tC_4H_9^+ + iC_5H_{12} &\rightarrow tC_4H_{10} + iC_5H_{11}^+
\end{aligned}
$$

though simple reactions can also proceed in this manner, for example

$$
H_2^+ + H_2 \rightarrow H_3^+ + H\cdot
$$

Most of these reactions show a transition at higher energies to the direct mechanism.

Many of these ion-molecule reactions have observed A factors which decrease with temperature rather than giving the expected increase. These can be explained by looking at the temperature dependence of the partition functions as used in transition state theory, Chapter 2.

INDEX

SYMMETRY AND GROUP THEORY IN CHEMISTRY

MARK LADD, Department of Chemistry, University of Surrey
Foreword by Professor The Lord Lewis, FRS, Warden, Robinson College, Cambridge

ISBN 1-898563-39-X   450 pages   1998

This clear, logical and up-to-date introduction is helpful to those reading any of the sciences wherein chemistry forms a significant part. It is copiously illustrated, including many stereoviews of molecules and hints on stereoviewing. The author's programs which aid the derivation, study and recognition of point groups are available on the Internet *www.horwood.net/publish*. **Worked examples, and problem exercises with tutorial solutions are given in extended tutorial form**.

Contents: Symmetry everywhere; Symmetry operations and symmetry elements; Group theory and point groups; Representations and character tables; Group theory and wave functions; Group theory and chemical bonding; Group theory, molecular vibrations and electron transitions; Group theory and crystal symmetry.

"Provides background to rationalize and synthesize the use of symmetry to problems in a wide range of chemical applications, and is a necessary part of any modern course."

Professor The Lord Lewis, FRS, Warden, Robinson College, Cambridge

"Treats the subject matter in a logical sequence, thoroughly and in depth, forming an excellent text for a major core second or third year course."

Dr. John Burgess, Reader in Inorganic Chemistry, University of Leicester

FUNDAMENTALS OF INORGANIC CHEMISTRY
An Introductory Text for Degree Course Studies

JACK BARRETT, Imperial College of Science, Technology & Medicine, London *and* MOUNIR A. MALATI, Mid-Kent College of Higher and Further Education, Chatham

ISBN: 1-898563-38-1   320 pages   1997

Contents: Introduction; Nuclear and radiochemistry; Electronic configurations and electronic states; Symmetry and group theory; Diatomic molecules and covalent bonding; Polyatomic molecules and metals; Ions in solids and solutions; Chemistry of s-block elements; Chemistry of *p*-block elements; Co-ordination compounds; Chemistry of *d*- and *f*-block elements.

"Constitutes a valuable addition to the ranks of introductory inorganic chemistry texts. This approach to electronic and bonding theory proves to be its strong point. It should indisputably be recommended to all who teach at this level and all departmental libraries."

Chemistry in Britain: J.Royal Society of Chemistry

"Undergraduates reading chemistry will find much benefit from these teachers' proper and kindly approach which will launch them into their more advanced part of the inorganic chemistry degree course. The book will be helpful also to those reading any of the sciences where chemistry forms a significant part".

Dr John Burgess, University of Leicester

IONS IN SOLUTION, Second Edition

JOHN BURGESS, Department of Chemistry, University of Leicester

ISBN: 1-898563-50-0   *ca.* 220 pages   1999

This up-to-date outline of the principles and chemical interactions in inorganic solution chemistry delivers a course module in an important area of considerable complexity which is only cursorily covered in most undergraduate texts.

Contents: Introduction; Solvation numbers; Ion-solvent distances; Ion-solvent interactions; Acid-base behaviour: hydrolysis and polymerisation; Stability constants; Redox potentials; Kinetics and thermodynamics; Kinetics and mechanisms: solvent exchange; Kinetics and mechanisms: complex formation; Kinetics and mechanisms: substitution at complexions; Kinetics and mechanisms: redox reactions; Past, present, future predictions.

REACTION MECHANISMS OF METAL COMPLEXES

R.W. HAY, School of Chemistry, University of St. Andrews

ISBN: 1-898563-41-1   *ca.* 200 pages   1999

This text provides a general background as a course module in the area of inorganic reaction mechanisms, suitable for advanced undergraduate and post-graduate study and/or research. The topic has important research applications in the metallurgical industry, and is of interest in the sciences of biochemistry, biology, organic, inorganic and bioinorganic chemistry.

Contents: Introduction to the field; The kinetic background; Substitution reactions of octahedral complexes; Substitution reactions in four- and five-coordinate complexes; Isomerisation and racemisation reactions; Reactions of coordinated ligands.

BIOINORGANIC CHEMISTRY, Second Revised Edition

R.F. HAY, Department of Chemistry, University of St Andrews

ISBN: 1-898563-45-4   *ca.* 250 pages   1999

This text provides a general background to the area of biological inorganic chemistry for final year and postgraduate courses and those beginning research in the field. The book has been modernised, restructured, updated and expanded. There is much new material and sections dealing with the functions of different metal ions (nickel, copper, vanadium, chromium, zinc), bio-mineralisation, non-metallic inorganic elements, inorganic drugs and mercury, aluminium, beryllium and chromium (VI).

"Probably the best yet produced for undergraduate teaching ... of value for those wishing to start bioinorganic research. I have no doubt that undergraduates and teachers will find it very useful."

Journal of Organometallic Chemistry (Dr. G.J. Leigh, University of Sussex)

Printed and bound by CPI Group (UK) Ltd, Croydon, CR0 4YY

03/10/2024

01040436-0017